Boston Studies in the Philosophy and History of Science

Volume 317

The series *Boston Studies in the Philosophy and History of Science* was conceived in the broadest framework of interdisciplinary and international concerns. Natural scientists, mathematicians, social scientists and philosophers have contributed to the series, as have historians and sociologists of science, linguists, psychologists, physicians, and literary critics.

The series has been able to include works by authors from many other countries around the world.

The editors believe that the history and philosophy of science should itself be scientific, self-consciously critical, humane as well as rational, sceptical and undogmatic while also receptive to discussion of first principles. One of the aims of Boston Studies, therefore, is to develop collaboration among scientists, historians and philosophers.

Boston Studies in the Philosophy and History of Science looks into and reflects on interactions between epistemological and historical dimensions in an effort to understand the scientific enterprise from every viewpoint.

More information about this series at http://www.springer.com/series/5710

Maria Cristina Amoretti · Nicla Vassallo
Editors

Meta-Philosophical Reflection on Feminist Philosophies of Science

Springer

Editors
Maria Cristina Amoretti
Department of Classics, Philosophy, and History
University of Genoa
Genova
Italy

Nicla Vassallo
Department of Classics, Philosophy, and History
University of Genoa
Genova
Italy

ISSN 0068-0346 ISSN 2214-7942 (electronic)
Boston Studies in the Philosophy and History of Science
ISBN 978-3-319-26346-5 ISBN 978-3-319-26348-9 (eBook)
DOI 10.1007/978-3-319-26348-9

Library of Congress Control Number: 2015954611

Springer Cham Heidelberg New York Dordrecht London

Printed on acid-free paper

Springer International Publishing AG Switzerland is part of Springer Science+Business Media
(www.springer.com)

Contents

1 **Some Key Topics in Feminist Philosophies of Science: An Introduction** 1
Maria Cristina Amoretti and Nicla Vassallo

2 **On the Possibility of Feminist Philosophy of Physics** 15
Maralee Harrell

3 **Climate Change Through the Lens of Feminist Philosophy** 35
Nancy Tuana

4 **Feminist and Non-feminist Philosophy of Biology: Parallels, Differences, and Prospects for Future Engagements** 55
Lynn Hankinson-Nelson

5 **Feminist Values, Commercial Values, and the Bias Paradox in Biomedical Research** 75
Kristen Intemann and Inmaculada de Melo-Martín

6 **Values and Evidence in Feminist Philosophy and in Neuroscience** 91
Robyn Bluhm

7 **The Reason/Emotion Divide in Contemporary Philosophy of Psychology** 113
Michelle Maiese

8 **Values in the Social Sciences: The Case of Feminist Research** 133
Kristina Rolin

9 **This is not a Manifesto: Archaeology and Feminism** 151
Pamela L. Geller

10 **Measuring the Value of Women: A Feminist Analysis of Economic Categories and Thought** 171
Ruth Hagengruber

11 The Woman of Reason: On the Re-appropriation of Rationality and the Enjoyment of Philosophy 185
Pieranna Garavaso

12 Feminist Versus General Philosophy of Science 203
Raffaella Campaner and Maria Carla Galavotti

Contributors

Maria Cristina Amoretti University of Genoa, Genova, Italy

Robyn Bluhm Michigan State University, East Lansing, MI, USA

Raffaella Campaner University of Bologna, Bologna, Italy

Inmaculada de Melo-Martín Weill Cornell Medical College, Cornell University, New York, NY, USA

Maria Carla Galavotti University of Bologna, Bologna, Italy

Pieranna Garavaso University of Minnesota, Morris, MN, USA

Pamela L. Geller University of Miami, Coral Gables, FL, USA

Ruth Hagengruber University of Paderborn, Paderborn, Germany

Lynn Hankinson-Nelson University of Washington, Seattle, WA, USA

Maralee Harrell Carnegie Mellon University, Pittsburgh, PA, USA

Kristen Intemann Montana State University, Bozeman, MT, USA

Michelle Maiese Emmanuel College, Boston, MA, USA

Kristina Rolin University of Helsinki, Helsinki, Finland

Nancy Tuana The Pennsylvania State University, University Park, PA, USA

Nicla Vassallo University of Genoa, Genova, Italy

Chapter 1
Some Key Topics in Feminist Philosophies of Science: An Introduction

Maria Cristina Amoretti and Nicla Vassallo

If we want to trace back the origins of feminist philosophies of science, we do not have to go too far. It is, in fact, only in the last few decades that feminist philosophers began to systematically look at the different sciences as possible objects of their inquiry. That being said, feminist philosophies of science are now flourishing fields characterized by lively and important discussions with regard to a rich and assorted collection of topics (see, e.g., collections like Alcoff and Potter 1993; Grasswick 2011; Harding 2008; Harding and Hintikka 1983; Jaggar 2008; Keller and Longino 1996; Kohlstedt and Longino 1997; Lennon and Whitford 1994; Pinnick et al. 2003; Vassallo 2009).

Even if we talk about "feminist" philosophies in opposition to "traditional," "mainstream" or "non-feminist" philosophies, we do not think that they stand for two independent and entirely separated sets of disciplines (see, e.g., Fricker and Hornsby 2000)—even if feminist philosophies all share an explicit concern about gender-related issues. In fact, one of the scopes of this collection is to emphasize and discuss both connections and differences between feminist and non-feminist philosophies of science, as well as to reveal viable and fruitful ways for the two to cooperate. This, of course, does not mean underestimating the significant new insights introduced by feminist philosophies of science.

Moreover, we prefer to talk about "philosophies" in the plural, as we think that there are many different ways to articulate a feminist philosophical approach to the sciences (see, e.g., Anderson 2011; Garavaso and Vassallo 2007; Potter 2007)—as the contributions of this volume also make clear. Still, some themes, methods, and problems are common to almost all feminist philosophies of the various sciences. Some of these issues have been introduced ex novo by feminist thinkers; others were already present in at least some non-feminist philosophies of science (see, e.g., Feyerabend 1975; Hanson 1958; Kuhn 1962; Quine 1969), but then were renewed, revised, and improved by feminist reflections. Some of these feminist threads have

M.C. Amoretti (✉) · N. Vassallo
University of Genoa, Genova, Italy
e-mail: cristina.amoretti@unige.it

N. Vassallo
e-mail: nicla.vassallo@unige.it

M.C. Amoretti and N. Vassallo (eds.), *Meta-Philosophical Reflection on Feminist Philosophies of Science*, Boston Studies in the Philosophy and History of Science 317, DOI 10.1007/978-3-319-26348-9_1

been acknowledged by non-feminist philosophies of science, and others are still regarded as largely unacceptable or at least suspicious (see Campaner, Galavotti this volume).

What is perhaps the most shared concept by feminist philosophies of science is that of a situated knower, and thus, of situated knowledge, which stands in harsh contrast to the "traditional" view of a disembodied and abstract knower, and thus, of transcendent knowledge or knowledge from nowhere (Amoretti and Vassallo 2011, 2012, 2013a; Code 1991; Haraway 1988; Harding 1991; Nelson 1990, this volume; Tuana this volume). This idea, which is not entirely absent from all non-feminist philosophies of science, has been developed by feminist thinkers in new ways by underlining the strong relations between gender and knowledge. Broadly speaking, our knowledge of the natural and social world depends on our social location; a location that is identified by different axes of social relations, one of which is gender (others being race, sexual orientation, ethnicity, occupation, etc.). This means that the gendered power structures of our society influence how and what we know (Harding 2008; Nelson 1993b).

Of course, advocating for the situatedness of knowledge does not compel one to accept epistemic relativism, nor to abandon the search for objective knowledge; however, it does raise new and important questions about the very notion of objectivity, which, of course, can no longer be conceived as transcendent and value-free (Antony 1993; Barwell 2007; Campaner, Galavotti this volume; Haely 2008; Haraway 1991; Nelson 1990). In rethinking objectivity, for example, one can argue that some social locations—typically those of marginalized and subordinate groups—are privileged or advantaged in determining what counts as knowledge, and thus, that they must be considered more objective than others (Harding 1986, 1991, 1993, 1995, 1998; Wylie 2003). Otherwise, objectivity can also be redefined in terms of intersubjectivity, being the result of critical discussion among a plurality of differently situated individuals and/or groups (Longino 1990, 2001).

Among the "traditional" features of objectivity that feminist philosophers typically criticize, there is also the subject/object dichotomy, which is the view that what is objectively real exists independently of knowers. As the social forces that shape subjects of knowledge are the same that also shape objects of knowledge, there can be no clear-cut distinction between subjects and objects. This kind of relationality, by eliminating the distance between subject and object, can lead to the development of less dominant interactions and new relational ontologies and methodologies (Duran 1991; Hrdy 1986; Keller 1983, 1985; Tuana this volume).

Another important view shared by many feminist philosophies, but that is also common to the naturalist turn in philosophy of science, is that of conceiving sciences not as bodies of theories but as sets of practices; more precisely, communal practices, characterized by complex interactions, both internal to specific communities and external to them, between different communities at various levels—from small laboratories to wider scientific communities (Antony 1993; Campaner, Galavotti this volume; Harrell this volume; Longino 1990, 2006; Nelson 1990, 1993a, 1995; Solomon 2001). That no epistemic subject in isolation is actually able to produce scientific knowledge independently from other epistemic subjects would be clear if we think about scientific discoveries: new scientific theories or hypotheses become knowledge only once they are tested, evaluated, and legitimated by a community of scientists.

Different communities, as epistemic agents, can, in principle, differ one from the other, share distinctive background assumptions (including theories, methods, ways of discourse, etc.), and produce different, or even contradictory, accounts of the natural and social world. However, communities, as epistemic agents, are internally more or less variegated and heterogeneous. In the feminist perspective, this represents a resource, because internal heterogeneity may be useful to generate new questions, uncover implicit biases, unveil hidden metaphors, and recognize different kinds of evidence. This means that introducing women and other underrepresented groups into a scientific community certainly helps to obtain scientific theories and hypotheses that are more reliable and objective, or, at least, less false (Hagengruber this volume; Longino 1990, 2001; Nelson 1990, 1993a, this volume; Solomon 2001).

The prevalent exclusion of women from the scientific enterprise is another hot topic in feminist philosophies of science. Women, in fact, have been long excluded or marginalized as both subjects and objects of scientific inquiry. On one hand, women are disqualified and disauthorized as knowers; because they cannot participate in the scientific enterprise or have more difficulty in doing so, their epistemic authority is denied (or at least diminished), they are deemed to have an essentially different cognitive style which, of course, is regarded as inferior to that of men (Bluhm this volume; Garavaso this volume; Maiese this volume). On the other hand, women are also disregarded as objects of knowledge, as they are not included in relevant studies (e.g., in medical trials), they are represented as inferior, deviant, or incomplete, and they are reckoned significant only as far as they effectively serve men's interests. Both these forms of exclusion, feminist philosophers argue, are not only morally but also epistemically negative and undesirable, and their consequences remain deleterious not only for women, but also for everybody else and, above all, for scientific knowledge in general (see, e.g., Amoretti and Vassallo 2013b, 2015; Anderson 1995a, b; Code 2004; Duran 1991; Fricker 2007; Geller this volume; Hagengruber this volume; Haraway 1991; Harrell this volume).

Another related theme that is central in the feminist philosophical discussion about the sciences is the critique of the value-free ideal—especially of its core: impartiality (see, e.g., Carrier et al. 2008; Kincaid et al. 2007; Lacey 1999). Many arguments have been advanced against it: some are endorsed by non-feminist philosophers as well. For instance, the argument from underdetermination of scientific theories, from pluralism with respect to epistemic values, from the pragmatic character of scientific inquiry, from inductive risk, from theory ladenness, and so on (see, e.g., Anderson 1995b; Intemann 2005; Longino 1990; Nelson 1993a; Putnam 1981, 2002). At the same time, the constitutive value-ladenness of science has been defended, stressing all the legitimate and beneficial influences of non-epistemic, or contextual, values (see, e.g., Anderson 2004; Bluhm this volume; Clough 2008; Harding 1998; Intemann 2001; Intemann, De Melo-Martín this volume; Longino

1979, 1983, 1994, 2001; Rolin this volume; Solomon 2001; Tuana this volume; Wylie and Nelson 2007).

However, those who support value-ladenness must face the problem not only of distinguishing between good and bad values, but also of showing that feminist values are among the good ones (see, e.g., Anderson 1995b, 2004; Antony 1993; Campbell 2001; Intemann, De Melo-Martín this volume; Longino 1990; Nelson 1990; Rolin this volume). Typically, good values are taken to be the ones that promote equality, pluralism, inclusion, democratization and progress, make gendered power relations visible, produce knowledge that is useful for people in subordinate and marginalized situations, minimize inequalities, and dismantle gender and other hierarchies. On the contrary, bad values create conditions of exclusion and marginalization, obstruct the emancipation of particular groups, slow or impede progress, hide gendered power relations, make certain subordinate and marginalized groups invisible and exclude them from knowledge, and promote stereotypes and hierarchies. In this respect, feminist values—by promoting inclusion, recognition, democratization, and equality, dismantling gender biases and hierarchies, exposing gendered metaphors, and supporting scientific theories that benefit egalitarian and liberation movements—should be certainly regarded as good values.

Indeed, feminist philosophies of science have always been particularly interested in and worried about the ethical, political, and social consequences of particular scientific hypotheses and theories, especially as far as life sciences and social sciences are concerned (see Code 2006; Harding 2004; Kourany 1998, 2010, 2012; Nelson this volume; Tuana this volume). In this sense, feminist philosophies of science are typically concerned with issues of justice and—contrary to non-feminist ones—may be often politically and socially engaged. However, it is worth stressing that, in their perspective, scientific theories that promote equality, inclusion, and liberation are not only ethical, political, and social advancements, but important epistemic improvements as well.

These and, of course, other pivotal issues that, perhaps in different ways, are part and parcel of both "traditional" philosophies of science and feminist philosophies of science are addressed in the various chapters of the volume with the intention to offer a sort of meta-philosophical reflection on the current status of feminist philosophies of science.

Whether most volumes dealing with feminist philosophies of science are quite general in nature, looking primarily at epistemological issues, the essays of this collection—using a range of different methodologies and styles, but all showing great clarity in both arguments and contents—systematically analyze feminist approaches and contributions to various philosophies of specific sciences; some of which have, so far, rarely been addressed by feminist philosophers. As far as we know, a similar focus on many distinct sciences all within one volume is rather unique. Each chapter is in fact explicitly devoted to a certain area of philosophy of science: philosophy of physics, of climate sciences, of biology, of biomedical sciences, of neurosciences, of psychology, of social sciences, of archeology, of economics, of logic, and, of course, general philosophy of science.

As said above, each essay uses distinct methodologies and styles to approach a certain field of philosophy of science: some essays present a critical and updated review of the feminist theories in the area under focus, making explicit what their innovative impact actually is (see Harrell; Tuana; Hankinson-Nelson); others aim to offer more specific and original contributions to current feminist positions (see de Melo-Martin and Intemann; Bluhm; Rolin), or try to find new parallels between feminist and non-feminist ideas (see Maiese); others focus on gender disparity and ostracism against women (see Geller; Hagengruber), or choose to deal with broader questions, such as rationality or the influence of feminist claims to the future course of the general philosophy of science (see Garavaso; Campaner and Galavotti). The following is a brief overview of the contributions.

To begin, in "On the Possibility of Feminist Philosophy of Physics," Maralee Harrell embraces the idea (common to the naturalistic turn in general philosophy of science), that physics can be better understood not as a body of theories expressed in static mathematical language, but as a dynamic activity performed by human beings, with their biases, interests, and values, in a complex and not-ready-made world. In this vein, physics can be thought of as a set of practices: more precisely, as a trichotomy of theoretical/methodological, social, and cognitive practices. Theoretical/methodological practices include the theories, methods, and background assumptions of physics, providing the foundations and metaphysical implications of physics itself; social practices involve the interrelations among groups of physicists as well as all the many group-level interaction dynamics. Finally, cognitive practices represent the "ways of thought," which might be idiosyncratic but mostly depend on internalized social norms used by individual physicists. According to Harrell, "traditional" and feminist philosophies of physics have quite different focuses on the discipline; if the former aims almost exclusively at investigating the theoretical/methodological practices of physics from a normative point of view, the latter is also interested in broadening the subject matter so to include the social and cognitive practices. Hence, feminist philosophy of physics has the great merit to engage in a critical analysis of the social and cognitive practices of physics. This analysis is not only important per se, but—as Harrell boldly argues—also allows for a better understanding of the theoretical/methodological practices themselves.

Discussing climate sciences, with her "Climate Change through the Lens of Feminist Philosophy," Nancy Tuana offers an overview of feminist philosophical reflections on anthropogenic climate change. In particular, she emphasizes that, unlike traditional philosophy of climate sciences, the lens of feminist philosophy has proven mainly successful in identifying some gendered conceptualizations in climate change knowledge, discourses, and practices that, despite being crucial, are often less visible than the differential impacts of climate change on the lives of individuals. These gendered constructions do, in fact, influence what we know, what we value, and how we act; thus, they are essential to promoting gender justice in the context of climate change. Tuana focuses on five relevant issues in the domain of climate change: the feminist critique of the ideal of a value-free science, which is essential to uncover gender biases in climate modeling and climate decision support science; the idea of positionality, which is central in evaluating matters of epistemic trust and/or distrust in climate change science; the notion of relationality, which leads us to consider new relational ontologies connecting humans and environments; the analysis of the discourses of climate change, which aims to unmask the interlink between women and less developed countries conceived as more vulnerable, passive, and poor; and the call for ecofeminism and care ethics in dealing with climate change. Given this, Tuana concludes that—by introducing new attention to the various domains of gender and to the interconnectedness of epistemological, ethical, ontological, and political issues—feminist philosophy makes original contributions to the reflections on anthropogenic climate change.

In the chapter dedicated to biology, "Feminist and Non-Feminist Philosophy of Biology: Parallels, Differences, and Prospects for Future Engagements," Lynn Hankinson-Nelson confronts non-feminist and feminist philosophy of biology in order to identify both substantive parallels and differences. On one hand, scholars of both fields engage issues of longstanding interest in the philosophy of science, but they focus more on actual scientific practices than on rational reconstruction. On the other hand, feminist philosophers of biology address topics, such as the relationship between sex/gender and biology, as well as areas and research programs, such as primatology or human evolutionary psychology, that are not typically pursued by "traditional" philosophers of biology; they are greatly concerned with the political, social, and/or ethical implications of biological hypotheses and are less interested in technical mathematical models. To see how much feminist philosophy of biology can differ from "traditional" philosophy of biology, but also can significantly enrich the discipline, Nelson focuses on three relevant themes: contextualism (the claim that scientists bring experiences reflecting their particular historical and cultural contexts to their own research); the presence of gender metaphors in biological theorizing and their consequences for biological knowledge; and the relationship between biology and political and/or ethical concerns. She then concludes that, given the growing interest in the general philosophy of science for promoting socially responsible science and engaging in the socially relevant and responsible philosophy of science, it would be possible for the feminist and "traditional" philosophies of biology to fruitfully engage one another, developing a cooperation that can and should be encouraged.

Moving to biomedical sciences, in their "Feminist Values, Commercial Values, and the Bias Paradox in Biomedical Research," Kristen Intemann and Inmaculada de Melo-Martín discuss the problem of commercial values, which is addressed in a very different way: with "traditional" and feminist philosophies of science. To begin, scholars of both parties agree that, in the context of biomedical research, commercial values pose relevant concerns about biases: they create conflicts of interest that can prejudicially affect scientific reasoning and have the power to establish research questions and priorities. However, if many "traditional" philosophers think that non-epistemic values of any kind should be avoided to preserve the impartiality of biomedical research, feminist theorists reject the ideal of impartiality, which is regarded as unattainable and undesirable. In this vein, feminist philosophers claim that feminist values ought to influence biomedical research because they can improve scientific inquiry and reinforce the concept of objectivity. Hence the "bias paradox:" if biomedical research should avoid commercial values because they are partial to non-epistemic interests, then—for the same reasons—it should avoid feminist values too; however, feminist values are defended. Intemann and de Melo-Martín argue that this paradox is misguided, as it is based on an equivocation about the sense of partiality. Of course, both commercial and feminist values are partial in promoting certain interests over others, but only commercial values are partial in arbitrarily excluding all other relevant values at stake. Quite differently, feminist values aim to reduce power inequalities and oppression, and thus, support the epistemic and social interests of all subjects.

In "Values and Evidence in Feminist Philosophy and in Neuroscience," Robyn Bluhm discusses the contribution of feminist theories to philosophy of neurosciences focusing on the conceptual framework of "brain organization theory," the view that purported cognitive, behavioral, and personality differences between men and women can be traced back to the brain, whose male or female structure is due to the exposure to different levels of sex hormones during fetal development. In her view, early feminist philosophers—emphasizing different aspects of this sex difference research and the context in which it was developed—have been much less critical of brain organization theory than were feminist scientists. This is due to their commitment to value-ladenness and empiricism. On one hand, believing that all science is shaped by values, early feminist philosophers could not simply claim that the presence of sexist values leads to bad science; on the other hand, endorsing empiricism, they could not rule out the possibility that research informed by sexist values was empirically well-grounded. In addition, their commitment to the idea that science is a social practice weakened their criticism of brain organization theory, which is not straightforwardly deemed as bad science. According to Bluhm, however, feminist philosophy can actually support a deeper critique of this neuroscientific conceptual framework. This can be done not by merely changing the social practices of neuroscientists but, instead, by developing new modes to critically assess the way that social issues and values about sex/gender permeate and shape empirical evidence and research.

According to Michelle Maiese's contribution, "The Reason/Emotion Divide in Contemporary Philosophy of Psychology," feminist philosophers of psychology can actually find an unsuspecting ally in some non-feminist philosophers, namely those working in the field of embodied and enactive cognition. This latter group of scholars aims especially at dismantling the reason/emotion divide, an assumption that permeates much of the work being done in contemporary philosophy of psychology. Roughly, this is the idea that reason and emotion are separate faculties of functions of the mind, which typically reflect several dichotomies such as active/passive, intellect/body, objective/subjective, and reflection/intuition. These dichotomies, in turn, have often been associated with binary constructions of gender: men are rational, active, intellectual, objective, reflective; women are emotional, passive, corporeal, subjective, intuitive. Moreover, as reason is generally viewed as the positive pole, while emotion as the negative one, the reason/emotion divide reinforces the idea that men are not only essentially different but also superior to women. Philosophers working in the field of embodied and enactive cognition typically argue that cognitive and rational processes are fully embodied and emotionally driven, and thus, that it makes no sense to speak of reason and emotion as if they were two different faculties of functions of the mind. On the contrary, they are constitutively interdependent and not clearly separable. According to Maiese, challenging the dichotomy of reason and emotion, these non-feminist philosophers make an important contribution, whether wittingly or not, to the dismantling of binary and essentialist ways of thinking about gender, and thus, to feminist philosophy of psychology.

Given that the idea of value-laden science is typically accepted by philosophers of social sciences, in "Values in the Social Sciences: The Case of Feminist Research," Kristina Rolin analyzes three models for understanding how feminist values can legitimately and constructively guide research in this very field, each of which is based on an argument against the value-free ideal. First, given the argument based on pluralism with respect to epistemic values, feminist values can interact with the epistemic values of empirical adequacy and explanatory power in order to establish what kind of evidence is relevant and what kind of causal information a certain explanation is expected to offer. Second, given the argument based on inductive risk, feminist values can decide what level of evidential warrant a theory is expected to meet. Third, given the argument based on value-laden background assumptions, feminist values can control the selection of those specific assumptions that are needed in evidential reasoning. These arguments, according to Rolin, show that feminist and epistemic values can fruitfully cooperate in the social sciences. Two alternatives to the value-free ideal—that is, Miriam Solomon's social empiricism and Helen Longino's critical contextual empiricism—are then presented and interpreted as attempts to integrate the liberal democratic values of equality and neutrality in science policy, that is, as both epistemic and political ideals. Given this interpretation, Rolin concludes that these approaches can give both epistemic and political justification for the view that feminist research projects have a legitimate place among other value-laden research projects in the social sciences.

The role of feminist values in guiding and shaping current scientific research is also discussed by Pamela Geller with regard to archeology. In "This is Not a Manifesto: Archaeology and Feminism," she explains that the political qualification of feminism has generated a certain ambivalence in archeology. On one hand, skeptics (among which there are also scholars who make gender focal in their studies of the past) claim that feminist values dangerously threaten the objectivity of archeology. On the other hand, feminist philosophers explain how feminist values can instead improve archeology; such values not only produce better practice in the present, but also deepen the understanding of past social organizations. Even if, generally speaking, feminist values produced a new commitment to the politics of recognition and inclusion, there are aspects of archeology that seem to resist any transformation. For example, demographic composition, which is still gender unbalanced in favor of men; acceptable research foci, which do not typically include gender-related investigations; and pedagogical emphases, which tend to avoid gender-related topics. To explain these resistances to feminist values, Geller explores the interconnections they have with power, focusing on the presence of both subtle (benevolent and not immediately experienced) and obvious (hostile and clearly experienced) sexism in present-day archeologist practices and institutional settings. From these reflections, she concludes that sexism pervades and persists not only because it is intersectional, contextual, and complex, but also because it includes structural, symbolic, and interpersonal violence, which may seem to be experienced idiosyncratically, but, in fact, is pervasive and exists collectively.

Moving to economics, Ruth Hagengruber's contribution, "Measuring the Value of Women: A Feminist Analysis of Economic Categories and Thought," aims to give an overview of the ways in which feminist philosophy of economics forcefully challenges the tradition of economic thought pointing to its ostracism against women, and also targets a transformation of economic categories. Mainstream economics sees itself as a value-free science that is committed to developing objective and universal methodological approaches in order to describe the self-interested actions performed by rational agents in a market. In so doing, however, it ignores that its own categories are constitutively embedded in specific cultural contexts and are influenced by social norms—among which there is the dichotomy between man/market and woman/house. Feminist philosophers of economics have shown that the market structure is not a pure, formal, non-corruptive, and inclusive institution, since it systematically ostracizes women, both as producers of economic output and as economic agents. One reason for this ostracism is that the market strongly depends on pre-market provisioning activities, which are ascribed to the house and, thus, to women. As women are systematically segregated from economic productivity, and their proficiency is relegated to care work, they cannot apply themselves as self-interested individuals. In this respect, feminist theorists not only ask for the inclusion of women in the marketplace, but also analyze the economic accountability of their domestic work, questioning if and how it can be put into an economically relevant determination. Hagengruber concludes, claiming that feminist philosophy of economics successfully contributes to revealing that gender disparity not only harms women, but everyone else as well.

The problem with determining the nature of reason and rationality, two central notions for sciences and philosophies of sciences, is specifically assessed by Pieranna Garavaso's contribution to feminist philosophy of logic, "The Woman of Reason: On the Re-Appropriation of Rationality and the Enjoyment of Philosophy." To begin, she explains why the definition of rationality is so relevant in a feminist perspective; as rationality is a central part of most philosophical notions of person, the exclusion of women from reason would mean to consider them only partial persons. Even appreciating current attempts to revise the "traditional" notion of rationality—that link it to emotions and stress its ethical relevance—Garavaso believes we must better understand what is wrong with it. Two general critiques are then considered; the former regards the notion of negation, the latter the adversarial method of argumentation. First, while Andrea Nye generally condemns logic, Val Plumwood specifically criticizes the standard notion of negation in classical logic, as it would represent an exclusive type of "otherness" that is unable to capture more nuanced forms of negation. Against this latter claim, Garavaso shows that Frege's notion of negation cannot be easily framed into the category of "otherness." Secondly, the adversarial method of argumentation in philosophy has been accused of obstructing feminist goals and being responsible for the small number of women philosophers. As Garavaso believes that this hypothesis is affected by essentialism, she proposes to endorse a more naturalistic conception of logic that is able not only to avoid essentialism, but also to support a feminist-friendly view of logic, rationality, and argumentation.

In the last contribution, "Feminist versus General Philosophy of Science," Maria Carla Galavotti and Raffaella Campaner critically examine key issues addressed by feminist epistemologies of the science. This analysis is twofold. On one hand, it stresses the convergences between feminist and current, but a non-feminist, general philosophy of science; on the other hand, it shows the exact original and specific contribution of feminist theories to the general philosophy of science. First, Campaner and Galavotti argue that a non-feminist philosophy of science is now better equipped to address the challenges posed by feminists (with regard to pluralism, the concept of objectivity, the link between cognitive and practical aspects of science, and contextualism), and, in fact, can eventually meet at least some of them by adopting a "bottom-up" approach to the discipline. Within this approach, the notions of explanation, prediction, and causality are, in fact, conceived in a pragmatic, pluralistic, and context-dependent way that goes hand in hand with a non-transcendent concept of objectivity. Second, the peculiar contributions of feminist theories—with regard to both the objects and to the methods of investigation—are examined in relation to three quite different fields: life sciences, chemistry, and criminology. To conclude, they evaluate whether and to what extent similar contributions can effectively influence the agenda and the future course of the general philosophy of science. Even if Campaner and Galavotti underline some problematic aspects that cannot be fully embraced, they also recognize the importance that all philosophers of science engage in adequate analyses of background assumptions and contextual social values, as well as of their relations with empirical evidence.

Acknowledgments We want to thank our contributors for their competence, cooperativeness, and patience, as well as an anonymous reviewer for his/her comments and suggestions. We also wish to thank Lucy Fleet for supporting and encouraging the project from the beginning and up to its completion.

References

Alcoff, L. M., & Potter, E. (Eds.). (1993). *Feminist epistemologies*. New York: Routledge.

Amoretti, M. C., & Vassallo, N. (2011). Scientific knowledge: Situatedness and intersubjectivity without standpoints. *Logic and Philosophy of Science, 9*(1), 517–522.

Amoretti, M. C., & Vassallo, N. (2012). On the independence of the social and situated dimension of scientific knowledge from the notion of standpoint. In S. Knauss, T. Wobbe, & G. Covi (Eds.), *Gendered ways of knowing in science: Scope and limitations* (pp. 57–74). Trento: FBK Press.

Amoretti, M. C., & Vassallo, N. (2013a). A way of saving normative epistemology? Scientific knowledge without standpoint theories. In V. Karakostas & D. Dieks (Eds.), *Epsa11 perspectives and foundational problems in philosophy of science* (pp. 467–478). Dordrecht: Springer.

Amoretti, M. C., & Vassallo, N. (2013b). Is there any problem with gender-specific medicine? *Verifiche, 42*(1–3), 139–156.

Amoretti, M. C., & Vassallo, N. (2015). Against sex and gender dualism in gender-specific medicine. In U. Mäki, I. Votsis, S. Ruphy & G. Schurz (Eds.), *Recent Developments in the Philosophy of Science: EPSA13 Helsinki* (pp. 357–367). Dordrecht: Springer.

Anderson, E. (1995a). The democratic university: The role of justice in the production of knowledge. *Social Philosophy and Policy, 12*, 186–219.

Anderson, E. (1995b). Knowledge, human interests, and objectivity in feminist epistemology. *Philosophical Topics, 23*, 27–58.

Anderson, E. (2004). Uses of value judgements in science: A general argument, with lessons from a case study of feminist research on divorce. *Hypatia, 19*, 1–24.

Anderson, E. (2011). Feminist epistemology and philosophy of science. In E. N. Zalta (Ed.), *The Stanford Encyclopedia of Philosophy*. http://plato.stanford.edu/entries/feminism-epistemology/. Accessed January 15, 2015.

Antony, L. (1993). Quine as feminist: The radical import of naturalized epistemology. In L. Antony & C. Witt (Eds.), *A mind of one's own: Feminist essays on reason and objectivity* (pp. 185–226). Boulder: Westview Press.

Barwell, I. (2007). Towards a defence of objectivity. In K. Lennon & M. Whitford (Eds.), *Knowing the difference. Feminist perspectives in epistemology* (pp. 79–84). London: Routledge.

Campbell, R. (2001). The bias paradox in feminist epistemology. In N. Tuana & S. Morgen (Eds.), *Engendering rationalities* (pp. 195–217). Albany: State University of New York Press.

Carrier, M., Howard, D., & Kourany, J. A. (Eds.). (2008). *The challenge of the social and the pressure of practice: Science and values revisited*. Pittsburgh, PA: University of Pittsburgh Press.

Clough, S. (2008). Solomon's empirical/non-empirical distinction and the proper place of values in science. *Perspectives on Science, 16*, 265–279.

Code, L. (1991). *What can she know? Feminist theory and the construction of knowledge*. Ithaca: Cornell University Press.

Code, L. (2004). The power of ignorance. *Philosophical Papers, 33*, 291–308.

Code, L. (2006). *Ecological thinking: The politics of epistemic location*. Oxford: Oxford University Press.

Duran, J. (1991). *Toward a feminist epistemology*. Savage, MD: Rowman & Littlefield.

Feyerabend, P. (1975). *Against method: Outline of an anarchistic theory of knowledge*. London: New Left Books.

Fricker, M. (2007). *Epistemic injustice: Power and the ethics of knowing*. Oxford: Oxford University Press.

Fricker, M., & Hornsby, J. (2000). Introduction. In M. Fricker & J. Hornsby (Eds.), *The cambridge companion to feminism in philosophy* (pp. 1–9). Cambridge: Cambridge University Press.

Garavaso, P., & Vassallo, N. (2007). *Filosofia delle donne*. Roma-Bari: Laterza.

Grasswick, H. (Ed.). (2011). *Feminist epistemology and philosophy of science: Power in knowledge*. London: Springer.

Haely, K. C. (2008). *Objectivity in the feminist philosophy of science*. London: Continuum.

Hanson, N. R. (1958). *Patterns of discovery: An inquiry into the conceptual foundations of science*. Cambridge: Cambridge University Press.

Haraway, D. J. (1988). Situated knowledges: The science question in feminism and the privilege of partial perspective. *Feminist Studies, 14*, 575–599.

Haraway, D. J. (1991). *Simians, cyborgs, and women: The reinvention of nature*. New York: Routledge.

Harding, S. (1986). *The science question in feminism*. Ithaca: Cornell University Press.

Harding, S. (1991). *Whose science? Whose knowledge? Thinking from women's lives*. Ithaca: Cornell University Press.
Harding, S. (1993). Rethinking standpoint epistemology: "What is strong objectivity"? In L. M. Alcoff & E. Potter (Eds.), *Feminist epistemologies* (pp. 49–82). New York: Routledge.
Harding, S. (1995). "Strong objectivity:" A response to the new objectivity question. *Synthese, 104*, 331–349.
Harding, S. (1998). *Is science multicultural? Postcolonialisms, feminisms, and epistemologies: Race, gender, and science*. Bloomington: Indiana University Press.
Harding, S. (2004). A socially relevant philosophy of science? Resources from standpoint theory's controversiality. *Hypatia, 19*, 25–47.
Harding, S. (Ed.). (2008). *Sciences from below: Feminism, postcolonialism, and modernities*. Durham: Duke University Press.
Harding, S., & Hintikka, M. B. (Eds.). (1983). *Discovering reality. Feminist perspectives on epistemology, metaphysics, methodology, and philosophy of science*. Dordrecht: Kluwer.
Hrdy, S. (1986). Empathy, polyandry, and the myth of the coy female. In R. Bleier (Ed.), *Feminist approaches to science*. New York: Pergamon.
Intemann, K. (2001). Science and values: Are value judgments always irrelevant to the justification of scientific claims? *Philosophy of Science, 68*, S506–S518.
Intemann, K. (2005). Feminism, underdetermination, and values in science. *Philosophy of Science, 72*, 1001–1012.
Jaggar, A. M. (Ed.). (2008). *Just methods: An interdisciplinary feminist reader*. Boulder: Paradigm Publishers.
Keller, E. F. (1983). *A feeling for the organism: The life and work of barbara mcclintock*. San Francisco: W.H. Freeman & Co.
Keller, E. F. (1985). *Reflections on gender and science*. New Haven: Yale University Press.
Keller, E. F., & Longino, H. E. (Eds.). (1996). *Feminism and science*. Oxford: Oxford University Press.
Kincaid, H., Dupré, J., & Wylie, A. (Eds.). (2007). *Value-free science? Ideals and illusions*. Oxford: Oxford University Press.
Kohlstedt, S. G., & Longino, H. E. (Eds.). (1997). *Women, gender, and science: New directions (Osiris, 12)*. Chicago: University of Chicago Press.
Kourany, J. A. (1998). A new program for philosophy of science, in many voices. In J. A. Kourany (Ed.), *Philosophy in a feminist voice. Critiques and reconstructions*. Princeton, NJ: Princeton University Press.
Kourany, J. A. (2010). *Philosophy of science after feminism*. Oxford: Oxford University Press.
Kourany, J. A. (2012). The ideal of socially responsible science: Reply to Dupré, Rolin, Solomon, and Giere. *Perspectives on Science, 20*, 344–352.
Kuhn, T. S. (1962). *The structure of scientific revolutions*. Chicago: University of Chicago Press.
Lacey, H. (1999). *Is science value free? Values and scientific understanding*. London: Routledge.
Lennon, K., & Whitford, M. (Eds.). (1994). *Knowing the difference. Feminist perspectives in epistemology*. London: Routledge.
Longino, H. E. (1979). Evidence and hypothesis: An analysis of evidential relations. *Philosophy of Science, 46*, 35–56.
Longino, H. E. (1983). Beyond "bad science": Skeptical reflections on the value-freedom of scientific inquiry. *Science, Technology and Human Values, 8*, 7–17.
Longino, H. E. (1990). *Science as social knowledge: Values and objectivity in scientific inquiry*. Princeton: Princeton University Press.
Longino, H. E. (1994). In search of feminist epistemology. *Monist, 77*, 472–485.
Longino, H. E. (2001). *The fate of knowledge*. Princeton: Princeton University Press.
Longino, H. E. (2006). The social dimensions of scientific knowledge. *The Stanford Encyclopedia of Philosophy*,
Nelson, L. H. (1990). *Who knows: From Quine to a feminist empiricism*. Philadelphia: Temple University Press.

Nelson, L. H. (1993a). Epistemological communities. In L. Alcoff & E. Potter (Eds.), *Feminist epistemologies*. New York: Routledge.

Nelson, L. H. (1993b). Who knows? What can they know? And when? *Reason Papers, 18*, 45–56.

Nelson, L. H. (1995). A feminist naturalized philosophy of science. *Synthese, 104*, 399–421.

Pinnick, C. L., Koertge, N., & Almeder, R. F. (2003). *Scrutinizing feminist epistemology: An examination of gender in science*. New Brunswick, NJ: Rutgers University Press.

Potter, E. (2007). *Feminism and philosophy of science. An introduction*. London: Routledge.

Putnam, H. (1981). *Reason, truth, and history*. Cambridge: Cambridge University Press.

Putnam, H. (2002). *The collapse of the fact/value dichotomy and other essays*. Cambridge, MA: Harvard University Press.

Quine, W. V. (1969). *Ontological relativity, and other essays*. New York: Columbia University Press.

Solomon, M. (2001). *Social empiricism*. Cambridge, MA: MIT Press.

Vassallo, N. (Ed.). (2009). *Donna m'apparve*. Torino: Codice Edizioni.

Wylie, A. (2003). Why stanpoint matters. In R. Figueroa & S. Harding (Eds.), *Science and other cultures: Issues in philosophies of science and technology* (pp. 26–48). New York: Routledge.

Wylie, A., & Nelson, L. H. (2007). Coming to terms with the values of science: Insights from feminist science studies scholarship. In H. Kincaid, J. Dupré, & A. Wylie (Eds.), *Value-free science? Ideals and illusion*. New York: Oxford University Press.

Chapter 2
On the Possibility of Feminist Philosophy of Physics

Maralee Harrell

2.1 Introduction

Physics is, like all sciences, not simply a collection of theories about how the world works. Theories and models of the world obviously play a critical role in physics, but they do not fully define the science. Rather, physics is an activity done by human beings with their attendant values, interests, and cognitive biases, in a complex, noisy world that does not come to us "carved at its joints." The dynamic nature of physics—the discovery, justification, negotiation, acceptance, and so on of its claims, models, experiments, and methods—cannot be captured through an exclusive focus on the static mathematical formulations of physical theories. Instead, we can more fruitfully think of physics, or arguably any science for that matter, as a set of practices, and more specifically, as a set of distinctively social, cognitive, and theoretical/methodological practices (Longino 1990).

Understanding science as practices, rather than theories or data, is certainly not novel to this chapter or even philosophy of physics. An emphasis on practice has been one of the most notable aspects of the recent "naturalistic turn" in general philosophy of science, in no small part due to the arguments of many feminist philosophers of science. Theories and models are important, but their presentation and structure often obscures key aspects of their history, motivation, and importance. Moreover, many of the most interesting activities in science take place outside of the traditional categories of theory and evidence. A major project of feminist philosophy of physics has been to shine a critical light on the social and cognitive practices in physics, and how those ultimately influence other aspects of the science.

M. Harrell (✉)
Carnegie Mellon University, Pittsburgh, PA, USA
e-mail: mharrell@cmu.edu

M.C. Amoretti and N. Vassallo (eds.), *Meta-Philosophical Reflection on Feminist Philosophies of Science*, Boston Studies in the Philosophy and History of Science 317, DOI 10.1007/978-3-319-26348-9_2

There are typically not sharp divisions between these three types of practices; for example, many of the social practices are presumably grounded in cognitive practices within each individual. Nonetheless, the trichotomy of social, cognitive, and theoretical/methodological (henceforth, just "TM") practices can provide a useful way to approach and discuss the practices of physics. The social practices of a science are the interaction patterns among groups of scientists. Inquiry about those practices thus involves questions about the systematic inclusion or exclusion of particular individuals or groups from the community, as well as the social behaviors and norms that allow certain individuals or groups to flourish once in the community, perhaps at the expense of others. The social practices also include the group-level interaction dynamics at all levels, from small-group collaborations, up through lab groups, departments, and even whole sub-domains of physicists (e.g., those working on quantum mechanics). We explore many of the more pernicious social practices in physics in Sect. 2.3.1.

The cognitive practices are the "ways of thought" employed by individual physicists. Some of these practices have only a local impact, as when an individual physicist has an idiosyncratic way of solving a problem. Many of the cognitive practices have a larger impact, however, often because they are essentially social norms that have been internalized, and so change the individual's thinking patterns. In particular, as we will see in Sect. 2.3.2, the cognitive practices in physics include many beliefs and expectations about the nature of knowledge, how it is constructed, and who has a privileged standing in the community. As a result, physicists interpret theories and data differently depending in part on whether the source conforms to those expectations. These cognitive practices obviously derive partially from the dominant social practices, but they have a local, individual impact directly on the physicists' ways of thinking about and understanding the world.

Finally, the TM practices of a science are the theories, background assumptions, experimental and analysis methods, and knowledge claims of that science. These practices are, in many ways, the focus of "traditional" philosophy of science and philosophy of physics. They provide the foundations and metaphysical implications of the science itself, as well as "how things work" in this particular science. Some of these TM practices have a logical or mathematical justification (e.g., using a consistent, unbiased statistical estimator); others are more historically grounded or contingent (e.g., in psychology, the p-value for a null hypothesis statistical test to indicate a "real" or "significant" effect is usually taken to be 0.05). Contemporary philosophy of physics—much like contemporary philosophy of chemistry, biology, or psychology—purports to study the foundations of the science, and as we see in Sect. 2.2, these foundations are, in practice, almost exclusively understood to be TM practices. Philosophers of physics see their job as engaging in a normative analysis of the TM practices. These analyses might be informed by descriptive findings, but typically only descriptions that outline the expected or permissible "moves" in the science. There is little impact of the detailed, culturally grounded descriptions of social and cognitive practices that are done by sociologists, anthropologists, and ethnographers of science. One of the key impacts of feminist philosophy of physics is precisely to shine a normative light on the social and cognitive practices of physics, which can thereby enable us to better understand and critique the TM practices themselves. But first, we begin by looking at the types of questions normally asked in philosophy of physics.

2.2 Contemporary Philosophy of Physics

The research areas of contemporary philosophy of physics can be divided roughly into three categories, although the work of some philosophers can fall onto more than one. The first is exploring the metaphysical implications of our current theories; the second is providing rigorous proofs concerning philosophically interesting topics; and the third is analyzing particular events in the history of physics that shed light on current philosophical topics.

2.2.1 Metaphysical Implications

The first area concerns the metaphysical implications of our current theories; that is, saying what it is that our theories are telling us about what the world is really like. One large sub-area of this research is into the correct, or at least, an understandable, interpretation of quantum mechanics. I explain this example in more detail than the others because it will facilitate the comparison with feminist philosophy of physics later.

There are well-established and mathematically equivalent formalisms that we can use to make predictions about quantum mechanical systems, and these predictions have been shown to be highly accurate. The mathematical formalism does not, however, provide an interpretation of the variables in the equations; the mathematical functions alone do not say how they map or correspond to parts of the world. Consider, for example, the time-dependent Schrödinger equation for a non-relativistic particle:

$$i\hbar\frac{\partial}{\partial t}\Psi(\mathbf{r},t) = \left[\frac{-\hbar^2}{2m}\nabla^2 + V(\mathbf{r},t)\right]\Psi(\mathbf{r},t) \tag{2.1}$$

where $\hbar$ is Planck's constant, $\mathbf{r}$ is the position in 3-dimensional space, and Ψ is the wave-function. This equation is supposed to govern all particles—like electrons, protons, and photons—that are the constituents of all of the physical matter in the universe. But how does this equation govern particles? Each particle has a different wave function associated with each of the different aspects of it that can be measured. The equation above describes a particle's position, but for any given particle, there is a wave function for its momentum, its polarization, its spin, etc.

At a purely mathematical level, each wave function characterizes each (corresponding) property of the particle, including the possible changes in that property over time. Let's use a simple example: electron spin. The electron isn't actually spinning, but it behaves as though it is spinning about an axis. We can measure the spin of an electron, and there are only two possible values: we will either measure it to be "spin-up" (i.e., as if it were spinning counter-clockwise looking down the axis) or measure it to be "spin-down" (spinning clockwise). The spin wave function for the electron tells us how probable it is that our measurement will come out "spin-up" or "spin-down." And in the particular case of electron spin, the corresponding wave function says that these probabilities do not (in the absence of a measurement) change over time. For example, if the electron has just been stripped from an atom the probability is 1/2, meaning that we have a 50 % chance of measuring it to be "spin-up" and a 50 % chance of measuring it to be "spin-down." If, however, we know the electron is "spin-up," then we have a 100 % chance of measuring it to be "spin-up" and a 0 % chance of measuring it to be "spin-down." More generally, if we take the absolute value of the square of the wave function (for position, momentum, spin, etc.) at any time, we will get a probability, for each of the possible states it could be in, of measuring the particle to be in that state.

Actually, since the electron behaves as though it is spinning along an axis, we can measure the spin of an electron along any dimension. If you think of the axis of the electron as being along the x-axis, our measurement of its spin will either be "spin-up$_x$" or "spin-down$_x$." We could instead obtain these measurements for the y-axis, the z-axis, and any axis in between. We cannot, however, measure the spin in along the x-axis *at the same time* we measure its spin along the z-axis. One problem that seems unique to, or at least amplified by, quantum mechanics is what happens when we perform these measurements in a series. Recall that if we measure the spin of the electron along the x-axis to be, for example, spin-up$_x$, then measuring it along the x-axis again is guaranteed to result in a measurement of spin-up$_x$. However, if we measure the spin along the x-axis to be spin-up$_x$, and then measure it along the z-axis, and then measure it along the x-axis again, there is only a 50 % chance that the result of the third measurement will be spin-up$_x$. It is as if we'd observed a person's hair color to be brown, and then observed that person's skin color, and then had no idea what the person's hair color would be if we observed it again!

So how do we make sense of this very strange mathematics that nonetheless gives the most accurate predictions of the world at the quantum level? Here's where philosophers of physics have stepped in.[1] The standard interpretation of quantum mechanics was developed primarily by the physicists who invented to it: Neils Bohr, Werner Heisenberg, and Max Born (Cushing 1994). This interpretation is the one that is presented in nearly all of the quantum mechanics textbooks for the past

[1]At the advent of the new quantum theory the philosophers of physics were the physicists themselves. In the past 50–60 years, however, interpreting quantum mechanics has shifted to those in philosophy departments, and "real" physicists have largely ceased to concern themselves with these issues.

century (Griffiths 2004; Messiah 1999; see Sakurai 1994). According to the standard interpretation, before the first measurement occurs in our scenario, the wave function of the particle is in a *superposition* of being spin-up$_x$ and being spin-down$_x$; that is, the particle is, in some sense, simultaneously in these two states. The oddness of superposition is popularly exemplified by Schrödinger's Cat that is (again, in some sense) simultaneously alive and dead. This superposition disappears once we make the measurement, however, as the wave function "collapses" to being just spin-up$_x$. At the same time, being in just spin-up$_x$ is the same as being in a superposition of being spin-up$_z$ and being spin-down$_z$. Thus, when we measure the spin along the z-axis, there is a 50 % chance of measuring it to be "spin-up$_z$" and a 50 % chance of measuring it to be "spin-down$_z$." Measuring the particle to be one or the other, say spin-up$_z$, "collapses" the spin wave function along the z-axis states, so the particle thereby goes back into a superposition of being spin-up$_x$ and being spin-down$_x$, so the chances of it being measured again to be spin-up$_x$ are back to 50 %. This idea that measurements collapse the wave function in a relatively "memory-less" way is absolutely central to the standard interpretation.

So, there are two questions for the standard interpretation: (1) what constitutes a measurement? And (2) what is the metaphysical status of the wave function? The first question is important because it seems now that the Schrödinger equation does not tell us how the wave function changes all the time: when there is a measurement, something happens to the wave function (the "collapse") that is not a part of the Schrödinger equation. Thus, on this interpretation, there is something very special about measurement and a complete theory should give a precise and detailed description of what constitutes a measurement.

The second question is important because, upon measurement, the wave function changes dramatically, and seemingly instantaneously. If the wave function is nothing more than a mathematical tool for producing probabilities or something that measures our knowledge, then there is no cause for concern. However, we have reason to think that the wave function is not just a conceptual tool, but rather is a real thing in the world that interacts with physical objects. First, as we just discussed, the apparatus for measuring the spin of the electron is itself physical, and the wave function seemingly changes because of an interaction with that physical thing. Second, postulating that the wave function can interfere with itself, like water waves can, explains the results of some very basic experiments with both light and particles (e.g., the one-slit and two-slit experiments). We initially thought we were talking about how just an electron behaves, but now it seems that we are talking about the behavior of both the particle and the wave function. Even worse, if the electron has a wave function for each of the possible measurements we could perform on it, then talking about a single particle actually means talking about a multitude, possibly an infinity, of different wave functions.

The standard interpretation is relatively silent about the status of the wave function and what constitutes a measurement, but many answers have been offered in different (non-standard) interpretations of quantum mechanics. Two of the most famous are from David Bohm and Hugh Everett. David Bohm's "pilot wave" theory (Bohm 1952; Cushing 1994) says there are two things—the wave and the particle—and the wave guides the particle like an ocean wave guides a surfer. Hugh Everett's "many-worlds" theory (Everett 1957; Maudlin 2011) says that when a measurement is made, the world branches into however many possible values the variable could have had, and that in each world a different value was obtained by the measurement. Other, less prominent interpretations include the "consistent histories" theory (Griffiths 2002; Omnes 1999), the "many minds" theory (Albert and Loewer 1988), the "relational" interpretation (Rovelli 1996), and the "Ithaca" interpretation (Mermin 1998). The core issue for philosophers of physics is how, if at all, to coherently understand and interpret the mathematics of quantum mechanics. That is, the focus is almost entirely on particular TM practices.

There are many other sub-areas of research into the metaphysical interpretation of our theories. Probability plays a key role in both quantum and statistical mechanics (i.e., the study of the behavior of large collections of particles). There are many different possible interpretations of probability, which can be roughly divided into objectivist and subjectivist. Interpretations of quantum mechanics often concern the ontological status of those probabilities. Probabilities also play an ambiguous role in the description of, and explanations for, the world given by statistical mechanics (Albert 2000; Ehrenfest and Ehrenfest 1959; Sklar 1995). Einstein's special and general theories of relativity also prompt a number of different questions. For example, there are issues about the real difference between inertial and non-inertial reference frames, and the equivalence of inertial mass and gravitational mass (Cassirer 1957; Earman 1995; Reichenbach 1958). There are also important epistemological and ontological questions concerning the nature and foundations of space and time themselves (Friedman 1986; Sklar 1977).

2.2.2 *Proofs*

The second broad area in philosophy of physics is providing rigorous proofs concerning philosophically interesting topics, such as time travel, determinism, probability, and the curvature of space. For example, David Malament has attempted to prove that simultaneity in special relativity is not conventional (Malament 1977; Sarkar and Stachel 1999); Jeffrey Bub and Rob Clifton proved that it is possible to associate a classical probability space with a quantum state (Bub and Clifton 1996; Bub et al. 2000); John Manchak proved that a time machine of they type proposed by John Earman, Christopher Smeenk, and Christian Wüthrich is actually possible (Earman et al. 2009; Manchak 2009); Jantzen (2011) proved that any permutation invariant theory is incompatible with a particle ontology; and Werndl (2009, 2011) proved that there are deterministic and indeterministic systems that produce the same observations. In some cases, these proofs focus on implications of the physics that are not necessarily of interest to "mainstream" physicists. In many other cases, however, these proofs blur the line between philosophy of physics and physics proper. In all cases, though, they contribute to our understanding of the TM practices.

2.2.3 *History and Philosophy of Physics*

A final broad area in philosophy of physics is the analysis of particular events in the history of physics that shed light on current philosophical topics like scientific explanation, confirmation, and progress. For example, Earman and Glymour (1980a, b) have argued that neither the 1919 observations of the eclipse nor the measurements of the red shift provided the unequivocal support for Einstein's theory of relativity, as so often has been assumed by both physicists and philosophers; Norton (2011) has used two case studies from the confirmation of two of Einstein's theories to draw general lessons about the material theory of induction; Jammer (1966) has produced a very thorough account of the development of Quantum Mechanics that provides an incredible insight into the actual practice of science, rather than the ahistorical treatments that scientific developments often receive in textbooks; and Torretti (2000) uses a survey of mathematical modeling by Galileo, Newton and others to support his skeptical view of the (current) standard cosmological model.

2.3 Feminist Philosophy of Physics

Feminist philosophy of physics encompasses many different questions, but often starts with the observation that women are highly under-represented in physics, even compared to other sciences. The dearth of women has been well-documented for decades (European Commission 2006; Megaw 1992), and is an obvious entry point for much feminist philosophy of physics. One can ask about the impact of the underrepresentation of women on the social, cognitive, and TM practices of physics. Perhaps more interestingly, one can try to determine the reasons for the relative absence of women in physics, and then ask how those underlying factors change or harm physics itself. In this section, we start by considering these issues with respect to the social, cognitive, and TM practices, but also use them as a launching point for other questions and challenges that emerge in feminist philosophy of physics.

2.3.1 *Social Practices*

In the first place, students in physics (as well as other sciences) are told that science is the way to rid oneself of prejudices, feelings, and particular viewpoints to discover the truth about the nature of an independent reality. What students actually get, however, is very different. Women, in particular, are often acutely aware that prejudices and feelings do not get checked at the door of the science classroom or lab. This is part of the "chilly" climate that is mentioned in many analyses of the lack of women in physics: sexist language, disparaging remarks, lack of encouragement, inappropriate allusions to one's gender in academic contexts, and outright sexual harassment (Rolin 2008). For example, when women are asked why there are so few women in science, they cite the unpleasantness of male dominated work places, both in the overt sexist behavior of men and the more subtle ways that women are ignored or sidelined (Ecklund et al. 2012).

In addition, there is a broad sociocultural perception among scientists that masculinity, in whatever form it manifests, is required to be a successful scientist, including success in physics (Urry 2008). For example, sociologist of science Merton (1957) tells his readers of the fierce competition to be recognized as the first one to discover some new phenomenon or theory. And prominent biologist Lewontin (1980: 186) claims that, "science is a form of competitive and aggressive activity, a contest of man against man that provides knowledge as a side-product."

In quite overt demonstrations of this assumption of masculinity, prominent women scientists is history have been described as being "manly." Dorothea Erxleben, one of the first women in Germany to earn a degree in medicine, was said to have "proven herself manly;" Emilie du Chatelet, the French translator of Newton's *Principia*, was praised by Voltaire as being a woman who was "a very great man;" and astrophysicist Cecilia Payne Gaposchkin was described by Edwin Hubble as "the best man at Harvard" (Bug 2000).

The identification of good science with masculinity even appears in other cultures. In *Beamtimes and Lifetimes*, Sharon Traweek compares the cultures of high-energy physics in the United States and Japan. The qualities that the Americans attach to the concept of a good, successful physicist are opposite to the qualities that the Japanese attach to the concept. In both cases, however, the qualities associated with the good physicist are the qualities that are taken to be masculine in the respective culture (Traweek 1988, 1992). American physicists value culturally masculine traits such as individualism, competition, bravado, and self-promotion. Japanese physicists, on the other hand, value quite different traits—cooperation, humility and inter-dependence—but they are again the ones that professional Japanese men, but not professional Japanese women, are taken to have (Traweek 1988).

There are often more subtle ways that women may feel unwelcome in physics, including the language and metaphors used by scientists to describe what science is. Nature is generally referred to as female, and often having secrets she doesn't want to give away. Merchant (1982) describes this powerful image of science as the "identification of nature with the female, especially a female harbouring secrets." Additionally, as Easlea (1986: 140) points out,

> Images of male scientists investigating female nature have abounded in modern science from the time of Francis Bacon onwards, ranging from quite ferocious images of putting nature on the rack and torturing her secrets from her, together with images of unveiling, piercing and particularly of penetrating the secrets of female nature, to gentle, loving images of male wooing of a wondrous, if coy, female nature.

The physicist, typically a man, speaks of science as a dance, a marriage, a seduction, or an assault (Harding 1986, 1991). For example, Richard Feynman, one hero of the physics community, says, "I am going to tell you what nature behaves like. If you simply admit that maybe she does behave like this, you will find her a delightful, entrancing thing" (Feynman 1965: 129). In 1984, high-energy physicist Frank Close explain to the readers of *The Guardian* that "[s]he [nature] has cleverly hidden some of her secrets from view in the macroscopic world. [...] By probing nature at the cutting edge may we have our eyes opened to her greater glory" (Easlea 1986).

Indeed, the image of science as man's quest to dominate nature has been widespread at least since Francis Bacon wrote *The Masculine Birth of Time* in the 16th century, but has been particularly salient in the physics community in the past hundred years. Since World War II, as American physicists became famous for developing the nuclear weapons that would end the war, physics has become very closely associated with the military (Easlea 1987). Dyson (1984) puts military scientists in the category of "nuclear warriors," and cites the competition between physicists as a major driving force behind nuclear proliferation. In the wake of the Second World War, strategists at RAND and the U.S. Air Force Strategic Air Command (SAC) competed to devise a strategy to attack the Soviet Union. Herman Kahn, one of the strategists, "coined the term 'wargasm' to describe the all-out 'orgiastic spasm of destruction' that the SAC generals supposedly favored" (Easlea 2002: 107).

If women and girls in the Western world are socialized from a young age to project "feminine" attributes such as caring, humility, and cooperation, then it is no wonder that they might feel out of place in a social arena that not only values masculine traits above feminine ones, but also encourages the domination of the feminine by the masculine to the point of orgiastic destruction. Indeed, research shows a decline in girls' participation in science, especially physics, starting in high school (Kahle 1987; Weinreich-Haste 1986). This research suggests that there is more going on than mere lack of interest (Danielsson 2010).

So, despite the fact that, when asked why there are so few women in science, *men* tend cite innate differences in scientific and mathematical ability (Ecklund et al. 2012), we need not postulate any cognitive differences to explain why so few women choose to become physicists. In fact, there is ample evidence that there are not any differences between men and women in the relevant cognitive abilities, such as mathematical cognition (Spelke 2005). But even if there are not relevant cognitive differences between men and women, we must still investigate the cognitive practices of physics, and consider how they may be hindering the success of women who do become physicists.

2.3.2 Cognitive Practices

Cognitive practices in science most obviously include the ways in which knowledge is constructed and achieved by particular scientists. Equally importantly, these cognitive practices also include the ways in which scientists *think* or *expect* knowledge to be constructed. These background beliefs about the processes and practices that can possibly lead to knowledge can be quite specific and stringent. As a result, many types of argument and investigation are declared simply irrelevant, and so people who prefer those modes of reasoning will be excluded or marginalized from the physics community. There is ample evidence that these cognitive practices, and expectations about those practices, are part of the explanation for the relative dearth of women in physics. Moreover, these expectations are particularly problematic because they *mis*characterize the ways in which physics (at least, modern-day physics) is actually done. Before considering the ways in which the expectations are mistaken, however, we first examine the cognitive practices and expectations themselves, with a particular focus on their impact on women's ability to enter and succeed in physics.

2.3.2.1 Who Can Be a Physicist? Who Can Construct Knowledge About Physics?

Historically, women have been systematically marginalized in physics, thereby creating biased expectations about who can contribute the production and construction of knowledge about physics. In fact, theories from physics were sometimes used (incorrectly, of course) to justify this exclusion. For example, according to physicist and educator Heinsohn (2000), some physicists in the late 19th century used the principle of energy conservation to argue that women could not both pursue a rigorous intellectual career and have strong, healthy babies. These arguments were ostensibly to prevent rivalry between men and women, but had the effect of making it harder for women to participate in the community of physicists. Similarly, according to Holland (2006), the foremost translator of 18th century German physicist Johann Wilhelm Ritter's scientific works, Ritter used concepts from the study of magnetism, namely polarization and indifference, to explain the "natural" longing women have to become pregnant and give birth. Even when women were able to participate in physics, their contributions were systematically devalued or relegated to mere technical improvements (Bug 2000; Shapin 1994).

These historical practices have been exacerbated by many features of physics pedagogy and the perception of physics in the wider culture, as these further reinforce the idea that only men can be successful physicists. Physics textbooks are notorious for giving the impression that the history of physics has been a linear development of great discoveries made by solitary gentlemen scientists. There are often boxed vignettes featuring a short biography and picture of the many famous (male) physicists including Galileo, Newton, Maxwell, Einstein, Bohr, and more recently Feynman (Danielsson 2010; Traweek 1988). The popular history of physics similarly focuses on the "great men" who are held to be responsible for modern-day physics. In fact, even small children nearly exclusively identify and portray scientists as male (Kahle 1987), with this effect diminishing only slightly as the child's age increases (Bug 2000). Unsurprisingly, people regard themselves as less able to contribute to an enterprise when the history of that enterprise does not include members of their group (Bug 2000).

All of these different factors lead to the expectation that successful physicists must be solitary, male geniuses, who construct new theories and knowledge by the sheer force of their own personal intellects. Many different types of individuals—most importantly (though not exclusively), women—are simply not expected sources of physics knowledge. The internalization of those cognitive expectations leads to increased (implicit) discrimination, reduced participation, and less chance for success (Schiebinger 2008). For example, studies show that both men and women rate the quality of a journal article higher when they believe the author is a man (Goldberg 1968; Paludi and Strayer 1985). Additional studies show that both women and men rate the quality of the curricula vitae of candidates for hiring or review higher if they believe that the candidate is a man (Foschi 2000; Steinpreis et al. 1999). One reason may be that women who succeed at jobs that have historically been male are judged to be less likable and have been more denigrated than similarly successful men (Heilman et al. 2004).

The internalization of these cognitive expectations not only excludes people from the domain of physics, but also can potentially eliminate various ways of knowing and knowledge construction. If knowledge is expected to be the product of a solitary genius, for example, then the results of collaboration will face a higher evidential and argumentative bar, simply because they do not conform to the dominant expectations. Moreover, these expectations not only lead to the exclusion and ignorance of women's contributions to physics, they also belie the reality that modern-day physics is actually practiced and constructed by communities of scientists working in close collaboration with each other.

2.3.2.2 Actual Practice Versus Expectations in Physics

As just noted, the dominant cognitive expectations about knowledge creators and constructors in physics is that they are solitary male investigators, engaged in the dispassionate, value-free quest for objective truths (Danielsson 2009; Keller 1985; Schiebinger 2008). As a result, people who violate these expectations—e.g., by being a woman, or passionate about the science, or working in a collaborative—are assumed to be (much) less likely to be producing or constructing knowledge. These pernicious effects are a significant harm on their own, but matters are made even worse because the expectations are actually false: the social, cognitive, and TM practices of modern-day physics simply do not conform to this model. Instead, modern-day physics involves passionate scientists working collaboratively to develop, articulate, and defend their preferred views using both evidential and value-laden grounds, all with the goal of constructing knowledge about the world. Physics is not a dispassionate mirroring of the world, but requires imagination and creativity (Easlea 2002). Indeed, research shows that physics teachers are explicit about the expectations of serious, dispassionate work, but nonetheless reward male behavior in the classroom and lab that is playful and imaginative (Hasse 2002).

One of the most important divergences between expectations and reality can be seen vividly through the lens of feminist philosophy of physics, inspired by broader observations in feminist philosophy of science. The standard belief is that physicists are engaged in the search for objective truths; for example, physics is understood as trying to find the "language" of the universe, or to establish truths that are independent of our particular, situated experience. There is thus no need for debate, discussion, or negotiation; we ought, on this picture, simply calculate (in our theories) and compare (with experimental results). One of the central insights of feminist philosophy of science has been the impossibility of such an Archimedean point that stands outside our own cognition and ways of knowing (Kitcher 2001). Claims that are held up as "objective" are actually "socially negotiated" (Longino 1990) and "partial and situated knowledge" (Haraway 1990: 183–202) that emerges only through the interactions of particular individuals with particular cultural, social, and cognitive values (Harding 1991).

In the context of physics, these observations reveal the importance of interaction and culture in the construction of knowledge. We cannot adopt a "view from nowhere," but rather must ask questions from our actual value-laden, goal-driven position. To use just one example, Robert Boyle's gas law and the science that surrounds it must be understood as emerging in part from the sociocultural milieu and circumstances in which he lived and investigated (Potter 1993, 2001; Shapin and Schaffer 1985). Moreover, the inevitable use of goals and values in science does not thereby make the science "bad" or wrong; matters are far more complex. A diversity of approaches and the need for negotiation in scientific contexts can lead to substantially better science. The particular challenge for women in physics is that these goals and values can be influenced by gender ideologies in ways that lead women to be less likely to join and contribute to the physics community (Rolin 1999).

Moreover, women who manage to enter physics suffer further from these divergences between expectations and reality. Women, in part simply by being women, do not conform to the dominant expectations about who can be a successful physicist, and so their research is systematically devalued. Women who do essentially the same work as men—both amount and quality—are perceived to be less competent and credible than their male counterparts (Traweek 1988), and receive less support from the broader community (Taylor 2010). In general, the results of their work is not trusted as much as if a man did the same work (Taylor 2010). These judgments even seep into female physicists' own self-perceptions, as their self-assessments undervalue their own work, relative to the work of male physicists (Correll 2001, 2004; Fiorentine and Cole 1992).

These biased, discriminatory cognitive practices—some employed by both men and women—are obviously closely related to the pernicious social practices from Sect. 2.3.1. Many of the cognitive practices can naturally be understood as internalized versions of the broader, more explicit social practices. And these cognitive practices have a similar impact on the science of physics: by excluding or devaluing the work of talented, capable members of particular group(s), the scientific community inevitably suffers in a multitude of ways. That being said, one might object that these practices only impede our progress in physics, rather than actually harming the science. That is, perhaps these practices, while socially and personally harmful, only lead to slower science, not worse science. This response is, however, based on wishful thinking about the possibility of the isolability of the TM practices from the social and cognitive ones, as we shall see in the next section.

2.3.3 Theoretical/Methodological Practices

Hardly anyone working in science or studying science today would challenge the claim that there is sex and gender bias in the work place and culture of physics; nonetheless, very few would say that this bias spills over into the actual results of physics. These kinds of biases may affect medicine, biology and other "soft" sciences in which people are the objects of research, but surely there can be no gender bias in the curvature of spacetime or the energy level of a particle, right? Urry (2008: 150), for example, argues that since photons (and other objects of study for physicists) are not gendered, "there is little freedom in—and certainly no gender-related influence on—the results of experiments or the interpretation of observations of the natural world." In addition, she says:

> The laws of physics simply know no gender. What gets studied has little or no relation to gender, with possible exceptions in applied areas. Here physics and astronomy and mathematics are very different from, say, archaeology or anthropology or biology, where gender is part of what is being studied (Urry 2008: 151).

But Urry's claims here are off the mark; as physicist Whitten (1996: 13) remarks, feminist science studies is *not* saying that her "scattering code produces different output because [she is] female". Rather, the claim is that gender can influence physicists' goals and values, and that we should thereby expect at least some (indirect) influence of gender on the results. This line of argument is pursued by Karen Barad, who offers a metaphysical interpretation of quantum mechanics that takes into account not only the views of the theory's creators, but also much recent research in feminist science studies (see, e.g., Barad 1995, 1996, 1999, 2007).

Like most philosophers of physics, Barad is concerned with understanding the metaphysical implications of our best physical theories. With quantum mechanics, she approaches this in several ways. She begins with Niels Bohr and his views about the nature of reality and the practice of science,[2] and builds an interpretation of quantum mechanics consistent with, but going much further beyond, these views. She explains how this interpretation is, on the one hand, realist and objective, and on the other hand, is inspired by and consistent with the general themes of feminist epistemology. Finally, she gives an account of one way that physics education

---

[2]As the point of Barad's account is not to provide a definitive interpretation of Bohr's philosophy, but only to use some of his well known views to create her own interpretation of quantum mechanics, I will not comment on the accuracy of her description of Bohr's views.

reinforces the misrepresentation of physics as a TM practice, which has the side effect of driving many people away from the discipline.

In most of her writing, Barad emphasizes the fact that the way the Heisenberg Uncertainty Principle is presented to students in lectures and textbooks is contrary to the way that Bohr conceived of the principle, and in fact contrary to the way Werner Heisenberg himself came to view it after discussions with Bohr (Murdoch 1987). The Uncertainty Principle, mathematically, is the relationship between the accuracy of the value of two complementary observables. Consider the position, x, and momentum, p, of a quantum particle. When we measure the particle's position, there is some error in that measurement, Δx. But then consider that, in measuring the position—by shining a light on it, for example—we necessarily change the particle's momentum by some amount. If we had a value for the momentum before, then that value has been changed (by the act of measuring the position) by Δp. The Uncertainty Principle says that no matter how hard we try, there is a limit on the amount we can reduce the error in either measurement, and that limit is given by the equation:

$$\Delta x \Delta p \geq \hbar/2 \tag{2.2}$$

Heisenberg's principle is usually explained by appealing to what we can *know* about the particle: if Δx is the uncertainty in our knowledge of its position and Δp is the uncertainty in our knowledge of its momentum, then our certainty about both quantities at the same time is limited by Heisenberg's equation.

However, this is not the right way to think about it, according to Bohr. Bohr thought a more accurate name for the principle would be the "Indefiniteness Principle" or something similar. His view is that there are certain sets of observables, like position and momentum, that are complementary, meaning not that the values cannot just both be *known* at the same time, but that the particle does not actually *have* definite values for both at the same time. The equation tells us that the more definite the value for position is, the more indefinite the value for momentum becomes, and vice versa. Thus, if a particle has an absolutely definite momentum, then the indefiniteness of its position would be infinite, meaning that the particle could literally be anywhere at all! Bohr and Heisenberg debated the epistemological and metaphysical interpretations of the Uncertainty Principle, until Bohr won Heisenberg over.

What this means is that measurement has a very prominent role in Bohr's interpretation of quantum mechanics, as we saw when we earlier discussed the standard interpretation. According to Bohr, all that we can know are the results of our measurements, and the measurements are merely interactions between a quantum system and a non-quantum (macroscopic) system. And not only is that all we can know, but that is all there is to know. In Bohr's view quantum mechanics is nothing other than an apparatus for making predictions about the outcomes of measurements. The outcomes of measurements are what Bohr calls "phenomena" in his later writings, where he insists that phenomena *are* reality.

To make this clearer, let's contrast this Bohrian view of science and experiment with the classical (Newtonian) view. In a classical (Newtonian) worldview, science is objective because

1. there is a clear distinction between the observer and the observed,
2. the objects of study have properties that are independent of the particular observer,
3. science aims to describe this observer-independent reality,
4. measurements are reproducible, and so free of bias, and
5. measurements are continuous and determinable, and so can be subtracted out when describing reality.

In contrast, on Bohr's view, science is objective because

1. the material apparatus required to make the measurements is what defines the concepts,
2. phenomena constitute reality,
3. measurements are reproductions of phenomena,
4. measurements are reproducible by the same configuration of material parts and experimental conditions, and
5. scientific theories describe this new conception of reality.

For Bohr, it isn't that an electron has a spin, and we send it through a measuring apparatus so that we can know what this spin is. Rather, for Bohr, it doesn't make sense to talk about any properties of particles in a context other than that of measurement. So what is real for Bohr is not that the electron is spin-up along the x-direction, for example; what is real is the phenomenon of our measuring apparatus interacting with the electron and producing the measurement spin-up along the x-direction. For Bohr, there is no observer-independent reality: reality is just the phenomena, and so is very much observer-dependent.

To describe this alternative conception of what is real, Barad introduces the term "agential reality," and describes her version of Bohr's philosophy of science as "agential realism." This new phrase is meant to convey a whole host of philosophical positions that are essentially feminist in their origin. She draws on the theories of Haraway (1990), Keller (1985), Harding (1986, 1991), Longino (1990), and Latour (1993) to identify the ways in which agential realism is a feminist philosophy of science. The most prominent ways in which this is true are the following:

> 1. Agential realism grounds and situates knowledge claims in local experiences: objectivity is literally embodied [...] objective knowledge is situated knowledge,
> 2. Agential realism privileges neither the material nor the cultural: the apparatus of bodily production is material-cultural, and so is agential reality [...] the apparatus [...] is not separable from phenomena,
> 3. Agential realism entails the interrogation of boundaries and critical reflexivity [...] phenomena are the embodiment of cultural practices within theory, and
> 4. Agential realism underlines the necessity of an ethics of knowing (Barad 1996: 179).

One of the most important aspects of agential realism concerns the notion of objectivity. Ordinary scientific realism contrasts objectivity with subjectivity. In the first place, objects of study are said to have objective properties, which are properties that the object possesses prior to any observation. These objects may also have subjective properties, which are the properties that a person attributes to the object upon interaction with the object. In the practice of science, "being objective" means setting aside or ignoring all together any subjective properties, and instead concentrating on discovering the objective properties of an object. The way to determine whether a scientist is being objective or has discovered an objective property is to ask whether another scientist can repeat that discovery (the observation, the experiment, etc.).

Agential realism dissolves the conceptual difference between objectivity and subjectivity with the claim that phenomena are the only reality. On this view, objects don't possess any properties that are independent of their interactions with other objects in the world. Rather, the *only* properties objects have are those that arise during these interactions. Thus, there can be no difference between objective and subjective properties. How, then, can we still claim that science is objective? The idea is that, under the same conditions—the same measuring apparatus, the same objects, and an observer—the observers will always witness the same phenomenon. As Barad (2007: 361) explains:

> In my agential realist account, scientific practices do not reveal what is already there; rather, what is "disclosed" is the effect of the intra-active engagements of our participation with/in and as part of the world's differential becoming. Which is not to say that humans are the condition of possibility for the existence of phenomena. Phenomena do not require cognizing minds for their existence; on the contrary, "minds" are themselves material phenomena that emerge through specific intra-actions. Phenomena are real material beings. What is made manifest through technoscientific practices is an expression of the objective existence of particular material phenomena. This is, after all, a realist conception of scientific practices. But unlike in traditional conceptions of realism, "objectivity" is not preexistence (in the ontological sense) or the preexistent made manifest to the cognitive mind (in the epistemological sense). Objectivity is a matter of accountability for what materializes, for what comes to be. It matters which cuts are enacted: different cuts enact different materialized becomings.

Ultimately, Barad takes herself to be engaged in the same kind of project as the philosophers of physics described in Sect. 2.2.1. She is presenting an argument "that agential realism can in fact be understood as a legitimate interpretation of quantum mechanics" (Barad 2007: 94). She is exploring the metaphysical implications of the mathematical formalism of quantum mechanics—i.e. trying to understand what quantum physics tells us about the way our world really works—so that we can improve our TM practices in physics.

2.4 Open Areas of Research in Feminist Philosophy of Physics

I want to close this chapter by indicating some areas of research in feminist philosophy of physics that are underdeveloped. The case of Barad's interpretation of quantum mechanics is the only one I know of that provides a feminist analysis of the TM practice of physics. The interpretation of quantum mechanics is an active area of research in philosophy of physics, and perhaps more feminist analysis could be done there. Other areas in which no one has ventured a feminist analysis are the interpretation of probability in physics, interpretations of special and general relativity, and theories of the foundations of space and time.

There are several feminist analyses of the cognitive practice of science, but analyses of knowledge production in physics are few. One excellent example, though, is Rolin (1999: 512) who argues that it is conceptually possible that gender ideologies influence the practice of physics, even "good" physics. In particular, she argues that gender ideologies can influence "how scientists understand or justify their cognitive goals or […] values." The question of whether gender ideologies have in fact influenced physicists in this way is an empirical one. Rolin describes two cases in which such influence is claimed, but finds the evidence lacking. There are a great many more cases, however, that have not yet been analyzed this way, and may provide some very fruitful research programs.

Finally, there are a wide variety of studies in the social practice of physics, but they are mainly confined to description rather than prescription. If feminist analyses are to have any real impact, we must go beyond describing the implicit and explicit sexism in physics and decide what changes to make in our pedagogical practices. As Auchinloss (1998: 15) notes, "feminist studies may hold a key to the success of efforts to attract and retain women […] create gender equitable environments […] and to reform physics education." Concrete proposals, based on sociological and psychological research, for improving physics education, as well as the testing of these proposals, are wide open areas of research.

Acknowledgments I would like to thank David Danks for extensive comments on earlier drafts.

References

Albert, D. Z. (2000). *Time and chance*. Cambridge, MA: Harvard University Press.

Albert, D., & Loewer, B. (1988). Interpreting the many worlds interpretation. *Synthese, 77*(2), 195–213.

Auchincloss, P. (1998). Physics and feminism. *APS News, 7*(5), 8.

Barad, K. (1995). A feminist approach to teaching quantum physics. In S. V. Rosser (Ed.), *Teaching the majority* (Vol. 20, pp. 369–375). New York: Teachers College Press.

Barad, K. (1996). Meeting the universe halfway: Realism and social constructivism without contradiction. In L. H. Nelson & J. Nelson (Eds.), *Feminism, science, and the philosophy of science* (pp. 161–194). Dordrecht: Springer.

Barad, K. (1999). Agential realism: Feminist interventions in understanding scientific practices. In M. Biagioli (Ed.), *The science studies reader* (pp. 1–11). New York: Routledge.
Barad, K. (2007). *Meeting the universe halfway*. Durham, NC: Duke University Press.
Bohm, D. (1952). A suggested interpretation of the quantum theory in terms of "hidden" variables, I and II. *Physical Review, 85*(2), 166–179, 180–193.
Bub, J., & Clifton, R. (1996). A uniqueness theorem for "no collapse" interpretations of quantum mechanics. *Studies in the History and Philosophy of Modern Physics, 27*(2), 181–219.
Bub, J., Clifton, R., & Goldstein, S. (2000). Revised proof of the uniqueness theorem for "no collapse" interpretations of quantum mechanics. *Studies in the History and Philosophy of Modern Physics, 31*(1), 95–98.
Bug, A. (2000). Gender and physical science: A hard look at a hard science. In J. Bart (Ed.), *Women succeeding in the sciences: Theories and practices across disciplines* (pp. 221–244). West Lafayette, IN: Purdue University Press.
Cassirer, E. (1957). *Substance and function and Einstein's theory of relativity*. New York: Dover Publications.
Commission, European. (2006). *She figures 2006*. Luxembourg: OPOCE.
Correll, S. J. (2001). Gender and the career choice process. *American Journal of Sociology, 106* (6), 1691–1730.
Correll, S. J. (2004). Constraints into preference. *American Sociological Review, 69*(1), 93–113.
Cushing, J. T. (1994). *Quantum mechanics: Historical contingency and the Copenhagen hegemony*. Chicago: University of Chicago Press.
Danielsson, A. T. (2009). *Doing physics—doing gender*. Uppsala: University Printers.
Danielsson, A. T. (2010). Gender in physics education research: A review and a look forward. In M. Blomqvist & E. Ehnsmyr (Eds.), *Never mind the gap!* (pp. 65–83). Uppsala: University Printers.
Dyson, F. (1984). *Weapons and hope*. New York: Harper & Row.
Earman, J. (1995). *Bangs, crunches, whimpers, and shrieks*. New York: Oxford University Press.
Earman, J., & Glymour, C. (1980a). Relativity and eclipses: The British eclipse expeditions of 1919 and their predecessors. *Historical Studies in the Physical Sciences, 11*(1), 49–85.
Earman, J., & Glymour, C. (1980b). The gravitational red shift as a test of general relativity. *Studies in the History and Philosophy of Science Part A, 11*(3), 175–214.
Earman, J., Smeenk, C., & Wuthrich, C. (2009). Do the laws of physics forbid the operation of time machines? *Synthese, 169*(1), 91–124.
Easlea, B. (1986). The masculine image of science with special reference to physics: How much does gender really matter? In J. Harding (Ed.), *Perspectives on gender and science* (pp. 132–158). New York: Falmer Press.
Easlea, B. (1987). *Fathering the unthinkable: Masculinity, scientists and the nuclear arms race*. London: Pluto Press.
Easlea, B. (2002). Patriarchy, scientists, and nuclear warriors. In J. A. Kourany (Ed.), *The gender of science* (pp. 98–112). New York: Pearson Education.
Ecklund, E. H., Lincoln, A. E., & Tansey, C. (2012). Gender segregation in elite academic science. *Gender and Society, 26*(5), 693–717.
Ehrenfest, P., & Ehrenfest, T. (1959). *The conceptual foundations of the statistical approach in mechanics*. New York: Cornell University Press.
Everett, H. (1957). "Relative state" formulation of quantum mechanics. *Reviews of Modern Physics, 29*(3), 454–462.
Feynman, R. (1965). *The character of physical law*. Cambridge, MA: MIT Press.
Fiorentine, R., & Cole, S. (1992). Why fewer women become physicians. *Sociological Forum, 7*(3), 469–496.
Foschi, M. (2000). Double standards for competence. *Annual Review of Sociology, 26*, 21–42.
Friedman, M. (1986). *Foundations of space-time theories: Relativistic physics and philosophy of science*. Princeton: Princeton University Press.
Goldberg, P. (1968). Are women prejudiced against women? *Trans-Action, 5*(5), 28–30.
Griffiths, R. B. (2002). *Consistent quantum theory*. Cambridge: Cambridge University Press.

Griffiths, D. J. (2004). *Introduction to quantum mechanics* (2nd ed.). New York: Prentice-Hall.
Haraway, D. (1990). *Simians, cyborgs, and women: The reinvention of nature*. New York: Routledge.
Harding, S. (1986). *The science question in feminism*. Ithaca, NY: Cornell University Press.
Harding, S. (1991). *Whose science? Whose knowledge?*. Ithaca, NY: Cornell University Press.
Hasse, C. (2002). Gender diversity in play with physics: The problem of premises for participation in activities. *Mind, Culture, and Activity, 9*(4), 250–269.
Heilman, M. E., Wallen, A. S., Fuchs, D., & Tamkins, M. M. (2004). Penalties for success. *Journal of Applied Psychology, 89*(3), 416–427.
Heinsohn, D. (2000). Thermodynamik und Geschlechterdynamik um 1900. *Feministische Studien, 18*(1), 52–68.
Holland, J. (2006). Die Zeit der Indifferenz. Johann Wilhelm Ritter und die Weiblichkeit. In S. Nieberle & E. Strowick (Eds.), *Narration und Geschlecht* (pp. 335–347). Cologne: Bölhau.
Jammer, M. (1966). *The conceptual development of quantum mechanics*. New York: McGraw Hill.
Jantzen, B. (2011). An awkward symmetry: The tension between particle ontologies and permutation invariance. *Philosophy of Science, 78*(1), 39–59.
Kahle, J. B. (1987). Images of science: The physicist and the cowboy. In B. J. Fraser & G. J. Giddings (Eds.), *Gender issues in science education*. Perth: Curtin University of Technology.
Keller, E. F. (1985). *Reflections on gender and science*. New Haven: Yale University Press.
Kitcher, P. (2001). *Science, truth, and democracy*. New York: Oxford University Press.
Latour, B. (1993). *We have never been modern*. Cambridge, MA: Harvard University Press.
Lewontin, R. (1980). "Honest Jim" Watson's big thriller about DNA. In G. S. Stent (Ed.), *James D. Watson, the double helix: A personal account of the discovery of the structure of DNA* (p. 186). New York: W.W. Norton & Company.
Longino, H. E. (1990). *Science as social knowledge*. Princeton: Princeton University Press.
Malament, D. (1977). Causal theories of time and the conventionality of simultaneity. *Nous, 11*(3), 293–300.
Manchak, J. B. (2009). On the existence of "time machines" in general relativity. *Philosophy of Science, 76*(5), 1020–1026.
Maudlin, T. (2011). *Quantum non-locality and relativity* (3rd ed.). Malden, MA: Wiley.
Megaw, W. J. (1992). Gender distribution in the world's physics departments. *Physics in Canada, 48*(1), 25–28.
Merchant, C. (1982). Isis' consciousness raised. *Isis, 73*(3), 398–409.
Mermin, N. D. (1998). The Ithaca interpretation of quantum mechanics. *Pramana, 51*(5), 549–565.
Merton, R. K. (1957). Priorities in scientific discovery. *American Sociological Review, 22*(6), 635–659.
Messiah, A. (1999). *Quantum mechanics*. New York: Dover Publications.
Murdoch, D. (1987). *Niels Bohr's philosophy of physics*. New York: Cambridge University Press.
Norton, J. D. (2011). History of science and the material theory of induction. *European Journal for the Philosophy of Science, 1*(1), 3–27.
Omnes, R. (1999). *Understanding quantum mechanics*. Princeton: Princeton University Press.
Paludi, M. A., & Strayer, L. A. (1985). What's in an author's name? *Sex Roles, 12*(3), 353–360.
Potter, E. (1993). Gender and epistemic negotiation. In L. Alcoff & E. Potter (Eds.), *Feminist epistemologies* (pp. 161–186). New York: Routledge.
Potter, E. (2001). *Gender and Boyle's law of gases*. Bloomington, IN: Indiana University Press.
Reichenbach, H. (1958). *The philosophy of space and time*. New York: Dover Publications.
Rolin, K. (1999). Can gender ideologies influence the practice of the physical sciences? *Perspectives on Science, 7*(4), 510–533.
Rolin, K. (2008). Gender and physics: Feminist philosophy and science education. *Science and Education, 17*(10), 1111–1125.

Rovelli, C. (1996). Relational quantum mechanics. *International Journal of Theoretical Physics, 35*(8), 1637–1678.
Sakurai, J. J. (1994). *Modern quantum mechanics*. New York: Addison-Wesley.
Sarkar, S., & Stachel, J. (1999). Did Malament prove the non-conventionality of simultaneity in the special theory of relativity? *Philosophy of Science, 66*(2), 208–220.
Schiebinger, L. (2008). Getting more women into science and engineering-knowledge issues. In L. Schiebinger (Ed.), *Gendered innovations in science and engineering* (pp. 1–21). Stanford: Stanford University Press.
Shapin, S. (1994). *A social history of truth: Civility and science in seventeenth-century England*. Chicago: University of Chicago Press.
Shapin, S., & Schaffer, S. (1985). *Leviathan and the air pump: Hobbes, Boyle, and the experimental life*. Princeton: Princeton University Press.
Sklar, L. (1977). *Space, time, and spacetime*. Oakland, CA: University of California Press.
Sklar, L. (1995). *Physics and chance: Philosophical issues in the foundations of statistical mechanics*. New York: Cambridge University Press.
Spelke, E. S. (2005). Sex differences in intrinsic aptitude for mathematics and science? *American Psychologist, 60*(9), 950–958.
Steinpreis, R. A., Anders, K. A., & Ritzke, D. (1999). The impact of gender on the review of the curricula vitae of job applicants and tenure candidates. *Sex Roles, 41*(7/8), 509–528.
Taylor, C. J. (2010). Occupational sex composition and the gendered availability or workplace support. *Gender and Society, 24*(2), 189–212.
Torretti, R. (2000). Spacetime models for the world. *Studies in the History and Philosophy of Modern Physics, 31*(2), 171–186.
Traweek, S. (1988). *Beamtimes and lifetimes*. Cambridge, MA: Harvard University Press.
Traweek, S. (1992). Big science and colonialist discourse: Building high-energy physics in Japan. In P. Galison & B. Hevly (Eds.), *Big science: The growth of large-scale research* (pp. 100–128). Stanford: Stanford University Press.
Urry, C. M. (2008). Are photons gendered? Women in physics and astronomy. In L. Schiebinger (Ed.), *Gendered innovations in science and engineering* (pp. 150–164). Stanford: Stanford University Press.
Weinreich-Haste, H. (1986). Brother sun, sister moon: Does rationality overcome a dualistic worldview? In J. Harding (Ed.), *Perspectives on gender and science* (pp. 113–131). New York: Falmer Press.
Werndl, C. (2009). Are deterministic descriptions and indeterministic descriptions observationally equivalent. *Studies in the History and Philosophy of Modern Physics, 40*(3), 232–242.
Werndl, C. (2011). On the observational equivalence of continuous-time deterministic and indeterministic descriptions. *European Journal for the Philosophy of Science, 1*(2), 193–225.
Whitten, B. (1996). What physics is fundamental physics? Feminist implications of physicists' debate over the superconducting supercollider. *NWSA Journal, 8*(2), 1–16.

Chapter 3
Climate Change Through the Lens of Feminist Philosophy

Nancy Tuana

The Fifth Assessment Report (AR5) of the Intergovernmental Panel on Climate Change (IPCC) underscored the widespread evidence of anthropocentric climate change. There are now direct observations of increases in global average air and ocean temperatures, widespread melting of snow and ice, and rising global average sea level (IPCC 2014, WG I). The Greenland and Antarctic ice sheets lost mass five to six times faster on average in the years 2002–2011 than the preceding decade. There are uncertainties about projections of future changes in the climate, but there is good reason to believe that should greenhouse gas emissions continue at or above the current rates, the negative impacts on human and natural systems will be high. Many countries and regions are already experiencing negative impacts on human well-being from changes in climate such as food and water insecurity from changing precipitation and climate-related extremes (IPCC 2014, WG II). These impacts will very likely accelerate over the next century. Climate change impacts all regions of the world, but it will have the greatest impacts on those countries and communities that are already struggling with poverty issues since it is expected to exacerbate poverty in developing countries as well as creating new poverty traps, particularly in countries with increasing economic inequality (IPCC 2014, WG II).

Based on the significance and severity of the impacts of human activities on the climate, some scientists and scholars have argued that we are entering a new era, the Anthropocene, in which humans have become "geological agents, changing the most basic physical processes of the earth" (Oreskes 2007: 73). They argue that "we need to fundamentally alter our relationship with the planet we inhabit" (Steffen et al. 2011: 739).

In the last decade a quickly growing body of literature written by feminist social scientists have revealed the complex ways in which the impacts of climate change differentially affect individuals and communities in various locations as result of inequalities based on gender, race, ethnicity, age, ability, and class (a small but representative sample includes: Dankleman 2010; Denton 2002; Masika 2002; Nagel 2015; Terry 2009). The United Nations Commission on the Status of

N. Tuana (✉)
The Pennsylvania State University, University Park, PA, USA
e-mail: ntuana@psu.edu

M.C. Amoretti and N. Vassallo (eds.), *Meta-Philosophical Reflection on Feminist Philosophies of Science*, Boston Studies in the Philosophy and History of Science 317, DOI 10.1007/978-3-319-26348-9_3

Women, for example, in 2002 called for action to "mainstream a gender perspective into ongoing research by, inter alia, the academic sector on the impact of climate change, natural hazards, disasters and related environmental vulnerability, including their root causes, and encourage the application of the results of this research in policies and programmes" (United Nations Commission on the Status of Women 2002: par. 7c). This important research has helped inform current understanding of the impact of climate change on socially and geographically disadvantaged people. The findings demonstrate that individuals and communities "exposed to persistent inequalities at the intersection of various dimensions of discrimination based on gender, age, race, class, caste, indigenity, and (dis)ability are particularly negatively affected by climate change and climate-related hazards" (IPCC 2014, WG II, 796). These studies reveal that climate change will very likely worsen existing poverty, exacerbate inequalities, and trigger new vulnerabilities as well as increase existing gender inequalities. Feminist inspired work on intersectionality is clearly at play in the IPCC representation of multidimensional vulnerability to climate related impacts (see Fig. 3.1).

Although feminist philosophical reflections on climate change have been significantly slower in being developed, feminist philosophical work in the fields of epistemology, philosophy of science, and environmental studies, as well as feminist ethical and social/political perspectives have the potential to make essential contributions to discourses on climate change. A feminist philosophical lens is, for

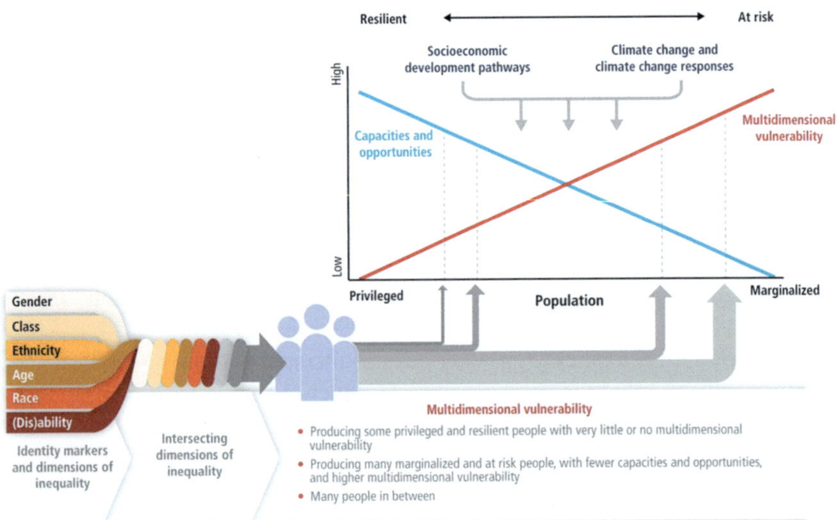

Fig. 3.1 Multidimensional vulnerability driven by intersecting dimensions of inequality, socioeconomic development pathways, and climate change and climate change responses. Vulnerability depends on the structures in society that trigger or perpetuate inequality and marginalization—not just income-poverty, location, or one dimension of inequality in itself, such as gender. (Reproduced from IPCC 2014, WG II, Fig. 13.5)

example, particularly salient in identifying gendered conceptualizations in climate change knowledge (sciences, including the social sciences) and practices (policy and activist responses). These gendered constructions are often less visible than the differential impacts of climate change on the lives of women and men. As, however, they permeate what we know and do not know, what we value highly and value less, as well as how we do and do not act, careful and critical attention to gendered constructions are essential to gender justice in the context of climate change. Feminist philosophical investigations across a range of questions and domains are essential for comprehending these gendered constructions and teasing out their ramifications. The special issue of *Hypatia: A Journal of Feminist Philosophy* (Tuana and Cuomo 2014) and its linked online symposium was designed to catalyze work in this field (see http://thephilosopherseye.com/phileye/online-events/hypatia-symposium-2).

3.1 The Value-Neutral Ideal in Science

The comparatively slow recognition of the gender-differentiated impacts of climate change is an example of the fact that what we do and do not know in the domain of climate change sciences has been affected by gender biases. Further, as MacGregor (2010: 124) has argued, "any attempt to tackle climate change that excludes a gender analysis will be insufficient, unjust, and therefore unsustainable." While MacGregor's call was for more feminist-informed sociological research into the ways in which the "material and discursive dimensions of climate change are gendered," her appeal applies equally to feminist philosophical analyses.

One important feminist philosophical lens for this work is feminist critique of the value-neutral ideal in science. Feminist philosophers of science have revealed the ways community-wide biases become entrenched in scientific research despite the pervasive belief in the objectivity of science and the view that science, when done properly, is not susceptible to interest, politics, or prejudice (Harding 1991; Keller and Longino 1996; Tuana 1989; Wylie 2002). While social biases about gender, class, or race are most likely to inform and limit scientific practice in the biological and social sciences (Fausto-Sterling 2000; Haraway 1997; Martin 1987), feminist scholarship has demonstrated that no field of science is immune from these impacts (Potter 2001; Traweek 1988). Scientific practice as a whole has also been the subject of careful feminist philosophical attention. Susan Bordo's analysis of "masculine ideal of objectivity" (Bordo 1987) and Genevieve Lloyd's groundbreaking study of "the man of reason" (1984) have revealed a gendered conception of reason at the heart of Western science. Sandra Harding's delineation of the interlinkages of its development with colonialization in *Sciences from Below: Feminisms, Postcolonialities, and Modernities* (2008), underscores the importance of a feminist lens not only for the *content* of science but also for its *methods*.

Feminist philosophical analyses of the role of values in climate modeling and climate decision support science is essential for rendering transparent the often ethical implications of those value choices. There is still much work to be done concerning the ways the values embedded in Integrated Assessment Modeling, as well as how the values that are often unconsciously incorporated into analyses of responses to climate change such as mitigation, adaptation, and geoengineering and their tradeoffs, can lead to injustices, including gender injustices.

To provide one example, consider the Intergovernmental Panel on Climate Change's (IPCC) demarcation of the central "reasons for concern" regarding climate impacts. Based on the commitment embedded in Article 2 of the United Nations Framework Convention on Climate Change to "prevent dangerous anthropogenic interference with the climate system" (UNFCCC 1992, Article 2), the IPCC's survey of peer reviewed literature identified five key vulnerabilities in their Third Assessment Report. These were: (1) Risks to unique and threatened systems; (2) Risks of extreme weather events; (3) Distribution of impacts; (4) Aggregate impacts; and (5) Risks of large scale discontinuities (McCarthy et al. 2001: Chap. 19). The colloquially named "burning embers" diagram was created to communicate the risks of rising global temperatures to a general public (see Fig. 3.2) and has been updated in the last two Assessment Reports. Each subsequent Assessment Report revealed that in all five key "reasons for concern" the temperature range that would likely avoid "dangerous anthropogenic interference" is getting smaller.

The impact of what feminists have labeled a "masculinized" conception of objectivity with its insistence on strict "neutrality" is reflected in the authors' discussion of the figure which emphasizes the objectivity of the IPCC findings.

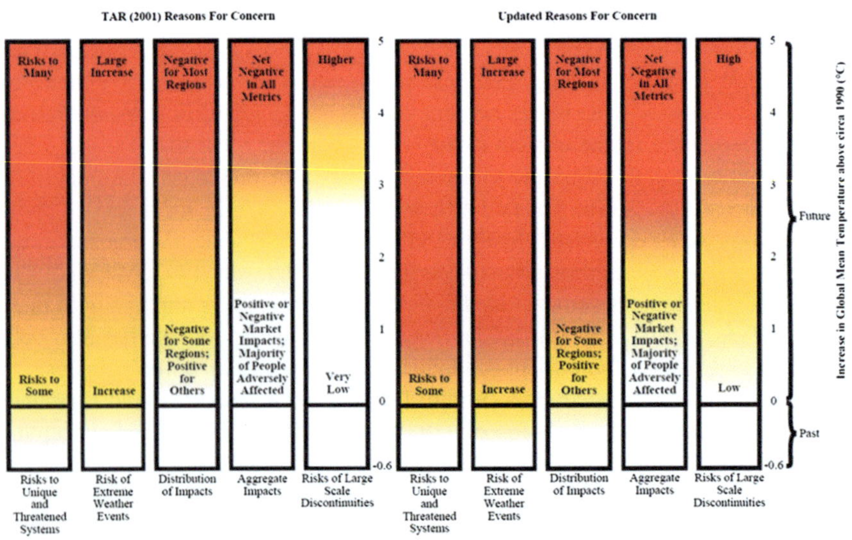

Fig. 3.2 Risks from climate change, by reason for concern on the left from *climate change 2001: impacts, adaptation and vulnerability* compared with updated data from Smith et al. (2009). Climate change consequences are plotted against increases in global mean temperature (°C) after 1990. (Reproduced from Smith et al. 2009)

> In presenting the 'embers' in the TAR [Third Assessment Report], IPCC authors *did not assess whether any single RFC [reason for concern] was more important than any other;* nor, as they noted, did they conclude what level of impact or what atmospheric concentrations of greenhouse gases would constitute DAI [dangerous anthropogenic interference], *a value judgment that would be policy prescriptive*. The 'embers' were designed primarily to communicate the associations of impacts with increases in GMT [global mean temperature] and facilitate examination of the underlying evidence for use by decision-makers contemplating responses to these concerns (Smith et al. 2009: 4133, my emphases).

Despite protests to the contrary, the selection of the five key reasons for concern involves a series of value judgments. Weighing harms and benefits against one another in itself requires value judgments about what constitutes a benefit or a harm and for whom. The mere selection and framing of the five key vulnerabilities—*these*—and the implicit assumption that these are all in fact key vulnerabilities—*these are all the most salient*—and that there are none missing from the list—*and no others*—all involve complex value judgments.

To illustrate this point consider the fourth reason for concern—*aggregate impacts*. The problem with "aggregate harms" as a measure is that it obscures the fact that relatively low aggregate harms is still compatible with very high levels of harm for some groups of people and for some ecosystems. Consider the threshold of 2° C as the target for limiting the average global surface temperature increase over the pre-industrial average identified by 2009 G8 Summit as the threshold for preventing dangerous anthropogenic interference with the climate system. In "Death by Degrees" (2009) Joni Seager analyzes the ideological framing of the 2° C rhetoric in a manner that reflects the problem of aggregation. Seager emphasizes that even with a 1° C warming there are "flora and fauna range shifts: increasing malaria (+300,000 deaths); extreme weather events […] severe food disruptions in the Sahel region of Africa" (Seager 2009: 16). She critiques the assumption that warming up to 2° C is "a degree of danger that is acceptable" (Seager 2009: 14), and reminds us that "many ecosystems and peoples will hit limits to adaptation long before 2° C, and some already have" (Seager 2009: 15). She demonstrates that for the millions of people in poor countries, low-latitude countries, low-lying states, and small island states, 2° C does not constitute "acceptable" levels of harm. In addition there are a number of ecosystems such as coral reefs where 2° C of warming will not prevent dangerous anthropogenic interference.

Aggregating harms may *appear* to be a value-free judgment. In fact, however, to embrace aggregation as an *acceptable* measure of harm presupposes the value-informed judgment that it is ethically and epistemically appropriate to have harmful and possibly devastating climate impacts from some regions be *offset* by the beneficial climate impacts other regions (see Tuana 2013). This is to impose a version of consequentialism where overall impacts determines the ethical acceptability of actions and elide the viewpoints of human rights or duty based ethical frameworks which would likely reject such aggregations as a violation of justice.

To unpack this issue even further requires consideration of *what was counted* as an aggregate impact and *how those aggregate impacts were measured*. Most studies that were used to create the burning embers figure measured loss of human life and economic impacts. But not all studies count loss of life years in the same way. Economic considerations often enter in with the result that loss of life years in poorer regions of the world are discounted via an economic measure that assigns lower values for the loss of life years in a poorer region of the world since individuals in these regions contribute a smaller amount to the GDP than individuals in wealthier regions (Broome 2004). Some economists urge that this calculation be rebalanced by weighting costs in poorer regions with income; but even this amendment embeds the value judgment that economic impacts (rather than psychological or aesthetic) are the appropriate measures. The problem with aggregating impacts is thus not only the issue of *whose* harms and benefits. The problem also includes *which* ones count most, e.g., income generating activities, and which harms and benefits do not count or count less, e.g., emotional well-being or ecosystem flourishing. These issues are in addition to who counts most, and who counts for less (Dietz et al. 2007; Nelson 2008, 2011; Stern 2007). Impacts such as loss of place for those who have to move due to rising waters or the emotions suffered when the places in which we live change so significantly due to droughts or floods that we no longer experience them as "home" (Tschakert et al. 2013; Tuana 2015a) are often ignored in such aggregative measures.

3.2 Who Knows?

A feminist philosophical analysis of the *reception* of knowledge is another important avenue of analysis. Particularly in countries like the United States where there has been often carefully cultivated distrust of climate change science (Oreskes and Conway 2010), the situatedness of the knower promises to be a fertile ground for analysis. Heidi Grasswick in "Climate Change Science and Responsible Trust: A Situated Approach" (2014) examines what constitutes responsible trust and/or distrust in climate change science. Deploying insights from feminist standpoint theory and epistemologies of ignorance (Tuana and Sullivan 2006) she examines how positionality is implicated in matters of epistemic trust. Grasswick (2014: 542) demonstrates that "a critical reflexivity of social position is required in order to produce knowledge well, and [...] argue[s] that this insight holds as well in the case of knowing through trust."

The work of feminist philosophers such as Code (1991, 2006) and Nelson (1990) provide a means to detect how the intricacies of one's particular situation shape knowledge production and reception. Code, for example, argues that subjective and objective factors are "constitutive of all knowledge worthy of the name" (Code 2006: 27). In other words, what an individual or a community views as reliable knowledge is shaped both by cultural and historical circumstances as well as by objective factors. Code concludes that it is important not only to consider the situatedness of the *content* of knowledge, but also to be attentive to the situatedness of the *knower.* Code's careful studies of women's epistemic situations, both in terms of their relationship to "expert knowledge" and their epistemic credibility in various contexts have led to extremely fruitful analyses of various forms of *epistemic injustice* caused by the situatedness of the knower (Fricker 2007; Medina 2012).

Adopting a situated approach, Grasswick argues for the value of and need for critical reflexivity aimed at understanding how those differently situated place their trust in climate change science. She argues that this critical reflexivity can be a valuable resource for appreciating and developing *responsible trust and/or distrust* in climate change science. Examining what has been labeled "the white-male effect," (Flynn et al. 1994) namely, the tendency for white males to perceive risks as much lower than do other demographic groups, Grasswick details the relevance of this phenomenon to understanding the dynamics of climate change belief and trust in the U.S. Studies on public perception of climate change reveal that American white males are, in comparison to other groups and genders, both less likely to be knowledgeable about climate change impacts and the least likely to endorse pro-environmental beliefs. Building on Elizabeth Anderson's study of criteria for judging the trustworthiness of climate change experts (Anderson 2011), Grasswick advocates a critical reflexivity, namely "an investigation into how one's positionality is shaping one's knowing (and one's ignorances)" (Grasswick 2014: 533) as a means to cultivating responsible trust in climate change science.

Lorraine Code in "Culpable Ignorance?" addresses a similar theme in her analysis of epistemic responsibility in relation to climate change skepticism and "the doubts that feed it and are nurtured to preserve it" (Code 2014: 670). Her goal is to address the

> politics-of-knowledge issues that are tangled in their implications for projects of developing reasoned, informed ways of countering climate change skepticism and addressing ecological ignorance, where countless members of affluent societies have much at stake in preserving the status quo with the alleged ignorance that sustains it, despite its ruinous consequences (Code 2014: 671).

Feminist and other liberatory philosophical work on sedimented ignorance is an important resource for understanding not only what we do and do not know about climate change (for example, the gender differential impacts of climate change), but also what we do and do not *believe* about climate change (for example, the reasons some do and some do not think it is anthropogenic), with careful attention to which "we's" are being invoked.

Feminist analyses must, of course, be developed through an intersectional lens to clarify the interlocking axes of difference relevant to knowledge practices. One domain of epistemic injustice concerns the perception of the role of traditional knowledge practices in the domain of climate change, both in terms of better understanding of the impacts of climate change as well as potential adaptative practices (Nakashima et al. 2012). Whyte's "Indigenous Women's Networks, Climate Change Impacts, and Collective Action" (2014), provides an illustration of distinctly indigenous efforts at climate change adaptation and mitigation. Whyte (2014: 600) emphasizes that some indigenous people see these efforts as crucial

> because climate variations can disrupt the continuance of the systems of responsibilities their community members self-consciously rely on for living lives closely connected to the earth and its many living, nonliving, and spiritual beings, like animal species and sacred places, and interconnected collectives, like forests and water systems.

Through a careful study of Anishinaabe women's responses to climate related water issues and what Whyte labels their "systems of responsibilities" arising from their situatedness, Whyte foregrounds the capacities for collective action unique to some indigenous women's groups, and argues that "they are forms of collective action that seek to engender global relations of coexistence *on indigenous people's terms*" (Whyte 2014: 612). Stressing the importance of recognizing and honoring indigenous women's traditional knowledge on mitigation and adaptation, Whyte (2014: 610) argues that "scientists and policy-makers have a political responsibility to include indigenous women's knowledge in their research, planning, and other empirical work" and to bolster the conditions needed for them to succeed.

The importance of critically engaging indigenous responses to climate change is the topic of Regina Cochrane's essay, "Climate Change, *Buen Vivir*, and the Dialectic of Enlightenment: Toward a Feminist Critical Philosophy of Climate Justice" (2014). Cochrane, while not advocating the wholesale acceptance of Latin American indigenous visions of *buen vivir*, or "living well," appreciates its inspiration for differently thinking relationality which is based on a relational ontology and co-related transformations in conceptions and practices of rationality. She highlights the parallels between the *buen vivir* attention to collectivity, harmony, reciprocity, complementarity, and interdependence among humans and with non-human nature and environmental feminist philosophy such as that of Val Plumwood. Deploying resources from Adorno and Horkheimer's critique of the European Enlightenment project, she argues for the development of "a feminist reconstruction of *buen vivir* and a critical feminist philosophy of climate justice [...] [that will assist in] reversing humanity's divorce from nature, promoting relocalization, and affirming deep ecology's expanded self" (Cochrane 2014: 591–2).

Attention to a wide and just inclusion of perspectives in climate responses is a central theme of Bronwyn Hayward's "Let's Talk about the Weather: Decentering Democratic Debate about Climate Change" (2008). Hayward offers an approach to resisting domination and building resilience inspired by Iris Marion Young's vision of decentered democratic processes. Advocating Young's vision which she sees as "primarily critical: to provide norms and criteria through which the normative legitimacy of the process and many of its policy outcomes can be questioned and improved" (2006: 48) Hayward illustrates the value of such an approach to ensure inclusion of a wide range of voices within planning and policy-making in regard to climate change in New Zealand and Pacific atoll communities.

3.3 Reconceptualizing What Is

In their investigation of Indigenous contributions to climate change responses, Cochrane and Whyte highlight a relational ontological perspective that is a component of such worldviews as well as an attentiveness to interlocking systems of responsibilities arising from relationality. The theme of relationality has been developed by Neimanis and Walker (2014) who argue that this attentiveness to relationality will require a new ontology. In "Weathering: Climate Change and the 'Thick Time' of Transcorporeality," they draw on feminist materialist theories to propose a new imaginary of climate change, namely, "weathering." Neimanis and Walker (2014: 559) propose "reconfiguring our spatial and temporal relations to the weather-world and cultivating an imaginary where our bodies are makers, transfer points, and sensors of "climate change." They see weathering as "a logic, a way of being/becoming, or a mode of affecting and differentiating that brings humans into relation with more-than-human weather [through which] we can grasp the transcorporeality of weathering as a spatial overlap of human bodies and weathery nature" (Neimanis and Walker 2014: 560). Neimanis and Walker argue that resisting dualisms between nature and culture or even the divide between climate change and our bodies will result in a different attitude to climate change. Rather than more "masculinist" attempts to control the weather they urge a "shift away from the 'stop climate change' temporal narrative [which] is not for us a weakening of possibilities for ethico-political engagement, but rather an opening up of a different sort of political and ethical orientation toward these questions: a politics of possibility and an ethics of responsivity" (Neimanis and Walker 2014: 561).

In "Being Affected by Climate Change: The Anthropocene and the Body of Ethics" (2015b) Nancy Tuana builds on the conception of viscous porosity that she initially developed in the wake of the 2005 hurricanes that devastated New Orleans. In "Viscous Porosity: Witnessing Katrina" Tuana (2008: 199–200) developed the contours of a relational ontology by calling attention to the

> viscous porosity of flesh—my flesh and the flesh of the world. This porosity is a hinge through which we are of and in the world. I refer to it as viscous, for there are membranes that affect the interactions. These membranes are of various types—skin and flesh, prejudgments and symbolic imaginaries, habits and embodiments. They serve as mediators of interaction.

In "Being Affected by Climate Change" she argues that an appreciation of the symbolic imaginary of the Anthropocene has the potential to affect an onto-ethical transformation of habits, affective dispositions, and ways of conceiving of the magnitude required to live differently in the face of anthropocentric climate change.

Tuana argues that the imaginary of the Anthropocene serves to catalyze a new ontological appreciation of the intimate, intricate, and inherent interrelations of what is which in turn dissolves the boundaries between humans/nonhumans, natural/cultural. She urges that we become ontologically affected by climate change through our appreciations of the co-constitution, both material and semiotic of "things in the making" (James 1958: 263). Through attunement to viscous porosity we come to appreciate the need for a new ethos for the Anthropocene, new ways of living together. Building on the conception of corporeal vulnerability advanced by Butler (2005, 2009) as the ability to affect and be affected, Tuana advocates for a shift from seeing vulnerability as linked to harm to an appreciation of the reciprocal vulnerability of things in the making as the ground of this ethos. Tuana argues that an Anthropocenean attention to corporeal vulnerability undertakes a transformation of the ethical by transforming the ability to be affected, to respond. Through this attunement we come to appreciate our interconnections with others, including past and distant others, and the intricate and irrevocable interconnections among beings and socionatural environments. We come to understand the dependency of future others on the lives we live today. She argues that we must "learn to be affected" (Latour 2004) by the full complexity of exchanges between human habits and geophysical interactions.

> We must learn to be affected by uncertainty and develop ways of knowing and living attuned to it. We must learn to be moved by, animated by, attuned to the threads of inextricable interconnections between consumption practices and ice sheets, between agricultural practices and species flourishing, between ocean currents and energy choices, between the way we live with the earth and the earth's becoming (Tuana 2015b).

3.4 Gendering the Discourse

Over the past decade there has been a growing body of philosophical literature on the topic of climate change with the majority of the publications focusing on philosophical analyses of climate ethics and climate justice.[1] However, a gender justice perspective is almost always absent from these accounts despite the fact, noted above, that there has been an explosion of work coming out of the social sciences on gender and climate change, and a vocal and visible activist movement whose aim is to convey the message that climate justice is inextricably interconnected with gender justice (Women's Environment and Development Organization, WEDO, http://www.wedo.org/; GenderCC, http://www.gendercc.net/). In the face of the growing number of philosophical books focusing on climate ethics or climate justice (see, for example, Arnold 2014; Broome 2012; Brown 2012; Coady and Corry 2013; Cripps 2013; Gardiner 2011; Gardiner et al. 2010; Moellendorf 2014; Shue 2014; Thompson and Bendik-Keymer 2012),[2] the wholesale absence of attention to gender in these studies suggests a form of *gender amnesia* too systematic

[1]There is some excellent work on philosophical analyses of climate models and the relevance of models for policy, (see, e.g., Diekmann and Zwart 2014; Parker 2010, 2011, 2013, 2014), as well as philosophical work on the science involved in solar radiation management, (see e.g., Tuana et al. 2012; Svoboda and Irvine 2014). However, with the exception of Buck et al. (2014) (discussed in this essay), none of this work engages a gender perspective.

[2]The exception to this pattern is Irwin (2010).

to be simply seen as a gap or omission. As Mills (1997) has argued with respect to race in Europe and the United States, there is an epistemology of ignorance functioning that involves an active production in which ignorance is preserved and linked to issues of cognitive authority, trust, credibility, and uncertainty. A "pattern of localized and global cognitive dysfunctions (which are psychologically and socially functional), producing the ironic outcome that whites [those in positions of power] will in general be unable to understand the world they themselves have made" (Mills 1997: 18). There is reason to suspect that an analogous epistemology of ignorance is at work with the gender amnesia currently limiting studies of climate ethics and climate justice. While perhaps not a conscious denial of the importance of the experiences and lives of women and other others in the context of a changing climate, there is, nonetheless, a persistent attention only to certain types of lives and experiences. This blind spot results in a failure to understand the full scope of climate justice issues in the very literature that aims to identify and rectify climate injustices.

While the inclusion of gender from an intersectional perspective is a vital contribution to work in the fields of climate ethics and climate justice, it also matters *how* gender is depicted. A gendered focus on climate change requires seeing the complex and intersectional interconnections between gender and vulnerability to climate impacts. While it is essential to include a gender dimension in scholarship on climate ethics, it is equally imperative not to frame women primarily as victims of climate change. While acknowledging the differential impacts of climate change on men and women, it would be a mistake to think that men are not negatively impacted by climate change due to gendered expectations and roles. Margaret Alston, to give just one example, examines long term drought in Australia and other aspects of climate variability. She shows that such weather events "together with lower socio-economic conditions and reduced farm production has combined to produce insidious impacts on the health of rural men" (Alston 2012) in Australia including higher levels of suicide. Furthermore, women's gendered roles can be a source of agency and resilience. There are a number of studies that demonstrate how women's gendered roles provide both resources and knowledges that make important contributions to adaptation. (See, for example, Glazebrook 2011.)

While it is essential to attend to the situated lives and experiences of women and men from an intersectional perspectives, it is equally crucial to apply a feminist philosophical lens to the *discourses* of climate change. The gendering of discourses often circulates in a complex fashion. Consider, for example, how discourses frame the so-called "Global North" and the "Global South." The "Global South" is often framed as more vulnerable, less powerful, less agential, and more in need of help from the "Global North." Tuana (2013) examines how this discursive trope circulates even in literature that explicitly addresses gender and climate change. Consider Fatma Denton's important and transformative early essay on gender and climate variability, "Climate change vulnerability, impacts, and adaptation: Why does gender matter?" (2002). How she framed her study was interlaced in a wider symbolic structure that was itself gendered. To give just one example from her text:

> The planet is a global concern incorporating a multitude of ecosystems, peoples, and cultures. As such, it requires collective input in its management, protection, and ultimately, its sustainability. Yet climate negotiations could be seen as a parody of an unequal world economy, *in which men, and the bigger nations, get to define the basis on which they participate and contribute to the reduction of growing environmental problems, while women, and smaller and poorer countries, look in from the outside, with virtually no power to change or influence the scope of the discussions* (Denton 2002: 10, my emphasis).

The italicized text illustrates the common discursive linkages between women and less developed countries (smaller and poorer countries) as less powerful, less capable of responding, having less agency, and ultimately more vulnerable. This feminization of the Global South often serves both to intensify the internalized depictions of women in the global south as passive, vulnerable, in need of another's agency, obscuring their resilience and agency, at the same time as it frames the Global South as a victim. While I used Denton to illustrate this point, it is essential to appreciate that this linkage is a widespread phenomena. It structures the imaginary of policies, for example, the Annex I (developed countries) and Annex II (less developed countries) framework of the Kyoto Protocol, as well as the repetitive images included in NGO publications and web pages which are well-intentionally designed to reveal the impact of climate change on women. In such venues the images of the women they focus on are located in what can clearly be read as a "less developed nation" and marked via their activities as poor, often rural, laborers (see Fig. 3.3). Through such common images and discourses, gender, race, class, and nation are interlinked with attributions of vulnerability, poverty, and passivity.

Fig. 3.3 The threats of climate change are not Gender-Neutral. (Reproduced from woman watch: http://www.un.org/womenwatch/feature/climate_change/)

Attention to these narratives is critical because they obscure the histories of "de-development" of the Global South by the Global North, overlooking the histories of colonialist and racist exploitation that have left the so-called "developing countries" with far fewer natural, economic, and social resources than would have otherwise been the case. As Cuomo (2011: 695) explains, "framing structural inequalities only in terms of susceptibility to harms focuses attention on the supposed weaknesses or limitations of those who are in harm's way, but says little about whether injustices or other harms have put them in such precarious positions." Such narratives also occlude the relevance of gender dynamics in the global North. Finally, such discursive tropes repeat a version of the "problem of aggregation" (previously discussed) in that that they assume a homogeneity of the "Global South" or of "women from the Global South."

Similar concerns have been raised regarding the concept of "vulnerable groups" and "vulnerable nations" in the discourses of climate change. Miller et al. (2010) argue that the labeling of certain groups or regions as vulnerable itself constitutes a type of stigmatization that can exacerbate marginalization and, as a result, undercut the agency and autonomy of those groups. The result is found in various forms of disempowerment. In an effort to address such concerns, Tschakert and Tuana (2013) argue for a resignification of both the concept of vulnerability and of resilience through the lens of a relational ontology. They propose the concept of "situated resilience" intimately linked to a transformed conception of vulnerability through the feminist lens of "corporeal vulnerability" found in the work of Judith Butler. They argue that a relational ontology entails an ethics of relation, one in which relationships between selves and others including nonhuman others are characterized by what they call "reciprocal vulnerability." Within a relational ontology, vulnerability is an "openness to the other through which each being's uniqueness emerges. Corporeal vulnerability is neither passive nor negative, but rather the ability to affect and be affected" (Tschakert and Tuana 2013: 86). It is the precondition for interrelationality, for care as well as injury, for benefit as well as harm. It thus occasions ethical obligations, such as attention to relationships of care and flourishing that are connected across space and time. Situated resilience proceeding from corporeal vulnerability "is attentive to the inseparable interactions between biophysical and socio-political dimensions of relationality […] It is not about invulnerability but about better ways of encouraging relations between peoples, both current and future, and between peoples and places" (Tschakert and Tuana 2013: 88–89).

The importance of feminist philosophical analyses of "the ways in which gendered environmental discourses frame and shape dominant understandings of the issue [of climate change]" (MacGregor 2010: 127) is a central concern of the work of Sherilyn MacGregor. Recognizing gender as a "discursive construction that shapes social life," she suggests that "gender analysis should involve the analysis of power relations between men and women and the discursive and cultural constructions of hegemonic masculinities and femininities that shape the way we interpret, debate, articulate and respond to social/natural/techno-scientific phenomena like climate change" (MacGregor 2010: 127). In her essay "Only Resist: Feminist Ecological Citizenship and the Post-Politics of Climate Change" (2014) MacGregor examines how the widespread control of climate discourses by men have produced the dominant scientific and policy framings for climate change that have resulted in an elitist discourse in which women are underrepresented and where, she worries, many will be alienated from participation. She illustrates how these same discourses delimit women's sphere to the private domain of household where, quoting Andrew Dobson, the goal is to change "light bulbs rather than changing society" (Dobson 2009).

> Will feminists—she worries—be able to express deep reservations about the assumptions being made about humans and nature in the scientized and securitized climate discourse? Resistance is difficult when the climate consensus has a tone of unquestionable scientific-moral authority [...] there is little room for ecofeminist critiques of scientific knowledge when an inaccessible, authoritative, and moralistic science sets the parameters of the issue (MacGregor 2014: 626–7).

She also raises concerns that the question of power has been largely ignored in climate discourses. She advocates for an ecofeminism that chooses the language of citizenship "over rhetorics rooted in fixed and private feminized identities that are themselves depoliticized (for example, maternalism)" (MacGregor 2014: 630). Her conclusion is that in the face of the control of climate discourses by the elite in ways that mute conflict and discourage dissent, there is a need for "agonal citizens" who will resist domination and articulate alternative visions and discourses.

3.5 Caring for Climate Change

MacGregor's call for an enhanced ecofeminism has been the goal of a number of feminist philosophers addressing the problem of climate change. In *The Ethics of Care: A Feminist approach to Human Security* (2011), Fiona Robinson applies feminist care ethics, an approach grounded on the view of the self as fundamentally relational, to develop a reconceptualization of human security. Her goal is "to consider how our view of security in global politics would change once we recognize and accept not just interdependence among states but the ways responsibilities and practices of care grow out of relations of dependence and vulnerability among people in the context of complex webs of relations of responsibility" (Robinson 2011: 4). As with many of the feminist philosophical analyses of climate change, Robinson embraces a relational ontology emphasizing that "the relational ontology of care ethics claims that relations of interdependence and dependence are a fundamental feature of our existence" (Robinson 2011: 5). While not supportive of essentialist tendencies of some ecofeminists, Robinson embraces a central tenet of ecofeminism, namely that environmental security is an integral part of human security. "Recognition of the mutual constitution of the human and natural worlds is a prerequisite to the development of an approach to human security based on the ethics and politics of care" (Robinson 2011: 143). Critical of the ways a "gender lens" can unwittingly reinscribe "stark dichotomies between men and women, thus obfuscating the relational nature of both men's and women's insecurity" (Robinson 2011: 157), Robinson works to develop what she argues is the broader lens of a critical ethics of care as a vehicle for addressing human security in the context of climate change. Chris Cuomo echoes these views arguing that "the cultivation of *responsible caring* attitudes and actions [is] necessary for the development of ethical and ecological relationships" (Cuomo 2011: 698). Both theorists emphasize the importance of critical attention to the socioeconomic injustices, past and present, out of which securities and insecurities are formed.

The feminist philosophically inspired lenses of ecofeminism and care ethics are applied by Buck et al. (2014) to a recent technological approach to dealing with climate change, namely, geoengineering. Geoengineering strategies are designed to deliberately alter the climate system in order to alleviate the effects of climate change. They are typically designed either to reduce the amount of absorbed solar energy in the climate system through what has come to be called Solar Radiation Management or to remove carbon dioxide from the atmosphere (Carbon Dioxide Removal). Holly Buck, Andrea Gammon, and Christopher Preston focus attention in "Gender and Geoengineering" on "where gender appears—and where it ought to appear—in the politics, ethics, and science of geoengineering" (Buck et al. 2014: 652). They identify four domains where gender is relevant: (1) the gender of those supporting geoengineering strategies; (2) the vision of control; (3) the design of the technologies; and (4) who it will most harm and benefit.

Focusing primarily on Solar Radiation Management, they note the demographics both of engineers and of policymakers involved with geoengineering strategies, and find that the vast majority are males. Moving from demographics to discourses and using an ecofeminst inspired lens, they identify the ways geoengineering solutions embed a masculinist approach that leads to "domination and objectification of the nonhuman world" (Buck et al. 2014: 655). Applying this ecofeminist lens to the field of engineering, they argue that geoengineering aligns with a culturally constructed masculine temperament "highlighting activity/dominance, objectivity, and technicality" (Buck et al. 2014: 657). Noting that issues of power, inequality, and domination are inextricably connected to the climate change problem, they note that the benefits of geoengineering will likely disproportionately accrue to developed countries while, at the same time, transfer environmental risk to what the Solar Radiation Management Governance Initiative (2011) labels "the poorest countries and the most vulnerable people." They advocate "value-sensitive design" as ways to ensure that technological responses to climate change attend to fundamental questions of justice and social transformation.

3.6 Conclusion

As this brief overview of feminist philosophical reflections makes clear, there is important work awaiting feminist philosophers in the domain of anthropogenic climate change. With feminist philosophical attention to the various domains of gender—from discourses to practices—and with appreciation for the intricate interconnectedness not only of peoples and places, humans and environments, but also for the important couplings of epistemological and ethical issues, ontological and political issues, feminist philosophical work in this field promises to transform both theory and practice in the face of climate change.

References

Alston, M. (2012). Rural male suicide in Australia. *Social Science and Medicine, 74*(4), 515–522.

Anderson, E. (2011). Democracy, public policy, and lay assessments of scientific testimony. *Episteme, 8*(2), 144–164.

Arnold, D. G. (Ed.). (2014). *The ethics of global climate change*. Cambridge: Cambridge University Press.

Bordo, S. (1987). *The flight to objectivity: Essays on Cartesianism and culture*. Albany: SUNY Press.

Broome, J. (2004). *Weighing lives*. Oxford: Oxford University Press.

Broome, J. (2012). *Climate matters: Ethics in a warming world*. New York: W.W. Norton and Company.

Brown, D. A. (2012). *Climate change ethics: Navigating the perfect moral storm*. New York: Routledge.

Buck, H. J., Gammon, A. R., & Preston, C. (2014). Gender and geoengineering. *Hypatia, 29*(1), 651–669.

Butler, J. (2005). *Giving an account of oneself*. New York: Fordham University Press.

Butler, J. (2009). *Frames of war: When is life grievable?*. London: Verso.

Coady, D., & Corry, R. (2013). *The climate change debate: An epistemic and ethical enquiry*. London: Palgrave Macmillan.

Cochrane, R. (2014). Climatic change, *buen vivir*, and the dialectic of enlightenment: Toward a feminist critical philosophy of climate justice. *Hypatia, 29*(1), 576–598.

Code, L. (1991). *What can she know? Feminist theory and the construction of knowledge*. Cornell: Cornell University Press.

Code, L. (2006). *Ecological thinking: The politics of epistemic location*. Oxford: Oxford University Press.

Code, L. (2014). Culpable ignorance? *Hypatia, 29*(1), 670–676.

Cripps, E. (2013). *Climate change and the moral agent: Individual duties in an interdependent world*. Oxford: Oxford University Press.

Cuomo, C. J. (2011). Climate change, vulnerability, and responsibility. *Hypatia, 26*(4), 690–714.

Dankelman, I. (2010). *Gender and climate change: An introduction*. London: Earthscan.

Deitz, S., Hope, C., Stern, N., & Zenghelis, D. (2007). Reflections on the Stern review: A robust case for strong action to reduce the risks of climate change. *World Economics, 8*(1), 121–168.

Denton, F. (2002). Climate change vulnerability, impacts, and adaptation: Why does gender matter? *Gender and Development, 10*(2), 9–20.

Diekmann, S., & Zwart, S. D. (2014). Modeling for fairness: A Rawlsian approach. *Studies in the History and Philosophy of Science, 46*, 46–53.

Dobson, A. (2009). 10:10 and the politics of climate change. *Open democracy*. http://www.opendemocracy.net/article/10-10-and-the-politics-of-climate-change. Accessed Jan 10, 2015.

Fausto-Sterling, A. (2000). *Sexing the body: Gender politics and the construction of sexuality*. New York: Basic Books.

Flynn, J., Slovic, P., & Mertz, C. K. (1994). Gender, race, and perception of environmental health risks. *Risk Analysis, 14*(6), 1101–1108.

Fricker, M. (2007). *Epistemic injustice: Power and the ethics of knowing*. Oxford: Oxford University Press.

Gardiner, S. M. (2011). *A perfect moral storm: The ethical challenge of climate change*. Oxford: Oxford University Press.

Gardiner, S. M., Caney, S., Jamieson, D., & Shue, H. (2010). *Climate ethics: Essential readings*. Oxford: Oxford University Press.

Glazebrook, P. (2011). Women and climate change: A case-study from Northeast Ghana. *Hypatia: A Journal of Feminist Philosophy, 26*(4), 762–782.

Grasswick, H. (2014). Climate change science and responsible trust: A situated approach. *Hypatia: A Journal of Feminist Philosophy, 29*(1), 541–557.

Haraway, D. (1997). *Modest_Witness@Second_Millenium.FemaleMan_Meets_OncoMouse*. New York: Routledge.

Harding, S. (1991). *Whose science? Whose knowledge? Thinking from women's lives*. Ithaca, NY: Cornell University Press.

Harding, S. (2008). *Sciences from below: Feminisms, postcolonialities, and modernities*. Durham, NC: Duke University Press.

Hayward, B. (2008). Let's talk about the weather: Decentering democratic debate about climate change. *Hypatia, 23*(3), 79–98.

IPCC. (2014). Summary for policymakers. In C. B. Field, V. R. Barros, D. J. Dokken, K. J. Mach, M. D. Mastrandrea, T. E. Bilir, M. Chatterjee, K. L. Ebi, Y. O. Estrada, R. C. Genova, B. Girma, E. S. Kissel, A. N. Levy, S. MacCracken, P. R. Mastrandrea, & L. L. White (Eds.), *Climate change 2014: Impacts, adaptation, and vulnerability*. Part A: Global and Sectorial Aspects. Contribution of Working Group II to the Fifth Assessment Report of the Intergovernmental Panel on Climate Change. Cambridge: Cambridge University Press.

Irwin, Ruth. (2010). *Climate change and philosophy: Transformational possibilities*. London: Continuum.

James, W. (1958). *Essays in radical empiricism: A plural universe*. New York: Longmans, Green.

Keller, E. F., & Longino, H. (1996). *Feminism and science*. Oxford: Oxford University Press.

Latour, B. (2004). How to talk about the body? The normative dimension of science studies. *Body and Society, 10*, 205–229.

Lloyd, G. (1984). *The man of reason: "Male" and "female" in western philosophy*. London: Methuen.

MacGregor, S. (2010). A stranger silence still: The need for feminist social research on climate change. *Sociological Review, 57*, 124–140.

MacGregor, S. (2014). Only resist: Feminist ecological citizenship and the post-politics of climate change. *Hypatia, 29*(1), 617–633.

Martin, E. (1987). *The woman in the body: A cultural analysis of reproduction*. Boston: Beacon Press.

Masika, R. (Ed.) (2002). *Gender and development: Special issue on climate change, 10*(2), 2–9.

McCarthy, J. J., Canziani, O. F. Leary, N. A., Dokken, D. J., & White, K. S. (2001). *Climate change 2001: Impacts, adaptation and vulnerability*. Contribution of Working Group II to the Third Assessment Report of the Intergovernmental Panel on Climate Change. Cambridge: Cambridge University Press.

Medina, J. (2012). *The epistemology of resistance: Gender and racial oppression, epistemic injustice, and resistant imaginations*. Oxford: Oxford University Press.

Miller, F., Osbahr, H., Boyd, E., Thomalla, F., Bharwani, S., Ziervogel, G., Walker, B., Birkmann, J., Van der Leeuw, S., Rockström, J., Hinkel, J., Downing, T., Folke, C., & Nelson, D. (2010). Resilience and vulnerability: complementary or conflicting concepts? *Ecology and Society 15* (3), 11. http://www.ecologyandsociety.org/vol15/iss3/art11/. Accessed Jan 10, 2015.

Mills, C. S. (1997). *The racial contract*. Ithaca, NY: Cornell University Press.

Moellendorf, D. (2014). *The moral challenge of dangerous climate change: Values, poverty, and policy*. Cambridge: Cambridge University Press.

Nagel, J. (2015). *Gender and climate change: Science, skepticism, security, and policy*. Boulder: Paradigm Publishers.

Nakashima, D. J., Galloway McLean, K., Thulstrup, H. D., Ramos Castillo, A., & Rubis, J. T. (2012). *Weathering uncertainty: Traditional knowledge for climate change assessment and adaptation*. Paris, Darwin: UNESCO, UNU.

Neimanis, A., & Walker, R. (2014). Weathering: Climate change and the 'thick time' of transcorporeality. *Hypatia: A Journal of Feminist Philosophy, 29*(1), 558–575.

Nelson, L. H. (1990). *Who knows? From Quine to a feminist empiricism*. Philadelphia: Temple University Press.

Nelson, J. A. (2008). Economists, value judgments, and climate change: A view from feminist economics. *Ecological Economics, 65*(3), 441–447.

Nelson, J. A. (2011). *Ethics and the economist: What climate change demands of us*. GDAE Working paper, Vol. 11(02).

Oreskes, N. (2007). The scientific consensus on climate change: How do we know that we're not wrong? In F. C. Dimento & P. Doughman (Eds.), *Climate change: What it means for us, our children, and our grandchildren* (pp. 65–99). Cambridge, MA: MIT Press.

Oreskes, N., & Conway, E. M. (2010). *Merchants of doubt: How a handful of scientists obscured the truth on issues from tobacco smoke to global warming*. New York: Bloomsbury Press.

Parker, W. (2010). Predicting weather and climate: Uncertainty, ensembles and probability. *Studies in History and Philosophy of Science Part B: Studies in History and Philosophy of Modern Physics, 41*(3), 263–272.

Parker, W. (2011). When climate models agree: The significance of robust model predictions. *Philosophy of Science, 78*(4), 579–600.

Parker, W. (2013). Ensemble modeling, uncertainty and robust predictions. *Wiley Interdisciplinary Reviews: Climate Change, 4*(3), 213–223.

Parker, W. (2014). Values and uncertainties in climate prediction, revisited. *Studies in History and Philosophy of Science Part A, 46*, 24–30.

Potter, E. (2001). *Gender and Boyle's laws of gases*. Bloomington: Indiana University Press.

Robinson, F. (2011). *The ethics of care: A feminist approach to human security*. Philadelphia: Temple University Press.

Seager, J. (2009). Death by degrees: Taking a feminist hard look at the 2° climate policy. *Kvinder, Køn, & Forskning, 3*, 11–21.

Shue, H. (2014). *Climate justice: Vulnerability and protection*. Oxford: Oxford University Press.

Smith, J. B., Schneider, S. H., Oppenheimer, M., Yohe, G. W., Hare, W., Mastrandrea, M. D., et al. (2009). Assessing dangerous climate change through an update of the Intergovernmental Panel on Climate Change (IPCC) "reasons for concern". *Proceedings of the National Academy of Sciences of the United States of America, 106*(11), 4133–4137.

Solar Radiation Management Governance Initiative. (2011). Solar radiation management: The governance of research. http://www.srmgi.org/files/2012/01/DES2391_SRMGI-report_web_11112.pdf. Accessed Jan 10, 2015.

Steffen, W., Persson, Å., Deutsch, L., Zalasiewicz, J., Williams, M., Richardson, K., et al. (2011). The anthropocene: From global change to planetary stewardship. *Ambio, 40*(7), 739–761.

Stern, N. (2007). *The economics of climate change: The Stern review*. Cambridge: Cambridge University Press.

Svoboda, T., & Irvine, P. (2014). Ethical and technical Challenges in compensating for harm due to solar radiation management geoengineering. *Ethics Policy & Environment, 17*(2), 157–174.

Terry, G. (Ed.). (2009). *Climate change and gender justice*. Oxford: Oxfam.

Thompson, A., & Bendik-Keymer, J. (2012). *Ethical adaptation to climate change: Human virtues of the future*. Cambridge, MA: MIT Press.

Traweek, S. (1988). *Beamtimes and lifetimes: The world of high energy physicists*. Harvard: Harvard University Press.

Tschakert, P., & Tuana, N. (2013). Situated resilience: Reframing vulnerability and security in the context of climate change. In J. Dugard, A. L. Clair, & S. Gloppen (Eds.), *Climate change talk: Rights, poverty and justice* (pp. 75–96). Claremont, South Africa: Juta Press.

Tschakert, P., Tutu, R., & Alcaro, A. (2013). Embodied experiences of environmental and climatic changes in landscapes of everyday life in Ghana. *Emotion Space and Society, 7*, 13–25.

Tuana, N. (1989). *Feminism and science*. Bloomington: Indiana University Press.

Tuana, N. (2008). Viscous porosity: Witnessing Katrina. In S. Alaimo & S. Hekman (Eds.), *Material feminisms* (pp. 188–213). Bloomington: Indiana University Press.

Tuana, N. (2013). Gendering climate knowledge for justice: Catalyzing a new research agenda. In M. Alston & K. Whittenbury (Eds.), *Research, action and policy: Addressing the gendered impacts of climate change* (pp. 17–31). Dordrecht: Springer.

Tuana, N. (2015a). Climate change and place: Delimiting cosmopolitanism. In J. Wahman, J. Medina, & J. Stuhr (Eds.), *Cosmopolitanism and place*. Bloomington: Indiana University Press.

Tuana, N. (2015b). Being affected by climate change: The anthropocene and the body of ethics. In K. Shockley & A. Light (Eds.), *Ethics and the anthropocene*. Cambridge, MA: MIT Press.

Tuana, N., & Cuomo, C. (Eds.) (2014). Special issue of *Hypatia* on climate change. *Hypatia, 29*(1).

Tuana, N., Sriver, R., Svoboda, T., Olson, R., Irvine, P. J., Haqq-Misra, J., et al. (2012). Towards integrated ethical and scientific analysis of geoengineering: A research agenda. *Ethics Policy and Environment, 15*(2), 1–22.

Tuana, N., & Sullivan, S. (2006). Introduction: Feminist epistemologies of ignorance. *Hypatia, 21* (3), 1–3.

United Nations Commission on the Status of Women. (2002). Report on the forty-sixth session (4–15 and 25 March 2002), Economic and Social Council. Official records 2002. Supplement No. 7 (E/2002/27-E/CN.6/2002/13). New York.

United Nations Framework Convention on Climate Change. (1992). http://unfccc.int/files/essential_background/background_publications_htmlpdf/application/pdf/conveng.pdf. Accessed Jan 10, 2015.

Whyte, K. P. (2014). Indigenous women's networks, climate change impacts, and collective action. *Hypatia, 29*(1), 599–616.

Wylie, A. (2002). *Thinking from things: Essays in the philosophy of archaeology*. Berkeley and Los Angeles: University of California Press.

Young, I. M. (2006). De-centering deliberative democracy. *Kettering Review, 24*(3), 43–60.

Chapter 4
Feminist and Non-feminist Philosophy of Biology: Parallels, Differences, and Prospects for Future Engagements

Lynn Hankinson-Nelson

4.1 Introduction: General Parallels and Differences

There are substantive parallels between feminist and non-feminist philosophy of biology. These include issues of interest, and some methodological and conceptual frameworks employed, in the research traditions (see also Fehr 2011). Both feminist and non-feminist philosophy of biology are interdisciplinary; they are engaged in by biologists and philosophers, often collaboratively. Although each tradition is dynamic and far from monolithic, some generalizations are appropriate.

In both, practitioners engage issues of longstanding interest in philosophy of science—including the nature of scientific theories and models, the role of laws in scientific theories, the nature of scientific explanation(s), the pros and cons of reductionism, and the role of epistemic values in scientific practice. Philosophers of biology frame these issues in relation to one or more of the biological sciences, rather than physics (a focus common in mid twentieth-century philosophy of science). As a result, the issues they emphasize are often different from those characterizing earlier philosophy of science.

Issues central to both feminist and non-feminist philosophy of biology include causal relationships in biological systems and processes, including whether all are uni-directional and "hierarchical," and whether biological generalizations are "law like;" the appropriateness of pluralistic explanations in specific biological fields or cases; conceptual issues specific to one or more biological sciences (e.g., in evolutionary biology, the nature of "fitness" and the level(s) at which selection occurs); and relationships between biology and its broader sociopolitical contexts—in particular how beliefs and assumptions in the historical and social contexts within which science is undertaken impact scientific theorizing, and vice versa. (However, as I will argue, until relatively recently, this focus has been more prominent in feminist phi-

L. Hankinson-Nelson (✉)
University of Washington, Seattle, WA, USA
e-mail: lynnhank@uw.edu

M.C. Amoretti and N. Vassallo (eds.), *Meta-Philosophical Reflection on Feminist Philosophies of Science*, Boston Studies in the Philosophy and History of Science 317, DOI 10.1007/978-3-319-26348-9_4

losophy of biology, particularly in terms of contemporary biology.) Philosophers of biology engage the pros and cons of reductionism, as did mid-twentieth century philosophers of science, but focus on issues specific to biology. Some analyze relationships between specific biological sciences (for example, the possibility of reducing explanations of classical genetics to those of molecular biology—e.g., Kitcher 1984); and reductionism in specific biological sciences (e.g., the rejection of group selection by some biologists and philosophers in favor of selection at the level of genes or individual organisms—e.g., Lloyd 2005; Sober and Wilson 1999).

In general, practitioners in the two traditions focus on *actual* scientific practices and, as a result, engage some issues that earlier philosophy of science did not engage. For example, in studying hypotheses, methods, and arguments that characterized Charles Darwin's reasoning to natural and sexual selection, many philosophers of biology do not limit their analyses to the then current *scientific* contexts (Darwin 1871). They study relationships between aspects of Darwin's reasoning, on the one hand, and aspects of the historical and social context of Victorian England that were not strictly scientific (or, in some cases, scientific at all), on the other hand (e.g., Ruse 2008). Nor, on the basis of the relationships their analyses reveal, do most conclude that Darwin's reasoning was "compromised."[1] Michael Ruse articulates a common attitude this way: "[I] do not mean to say that only culture counts and the facts go for naught. That is just silly. [...] We shall see again and again that the empirical evidence drives claims and counterclaims." At the same time, Ruse notes, "[T]o think that science stands apart from the rest of life, pure and disinterested, is just as silly" (Ruse 2013: ix–xi). (Although, again, as I argue below, until relatively recently and with some notable exceptions, non-feminist philosophers of biology have not engaged such issues in their analyses of *contemporary* biology to the extent that feminists have.)

This methodological-epistemological approach stands in sharp contrast to the efforts of earlier philosophers of science "to rationally reconstruct" scientific reasoning. Proponents of such reconstructions held that the ways in which a hypothesis or theory are generated (including any relationships to beliefs in the larger social context and/or to social factors characterizing scientific communities) are unimportant. For example, Carl Hempel, among others, took the nature of the justification of scientific theories to be the core issue of philosophy of science. This led to the development of formal schemas of the testing of hypotheses and theories, and to ignoring the details and contexts of the discovery and formulation of theories. Many contemporary philosophers of biology view accounts of "scientific reasoning" that only focus on justification as at best incomplete. This is not to say that philosophers of biology do not analyze formal models of biological processes that biologists develop, or develop formal models to characterize reasoning in biology. It is to say that, in general, they are more interested in studying actual scientific reasoning than they are in rationally reconstructing it.

Fehr (2011) has suggested that the parallels between feminist and non-feminist philosophies of biology suggest that these traditions could constructively engage one

[1]There are exceptions. See, e.g., Hubbard (1983).

another—a possibility that I explore in my conclusion. I first discuss some significant differences in the theoretical emphases of the two traditions. Feminist philosophers of biology engage issues and questions that, with some exceptions, are not regularly pursued by their non-feminist colleagues. Most obviously, feminists are concerned to understand the complex and multi-faceted relationships between *gender* and biology. They have sought to identify the multi-faceted ways in which biological research is often informed by unwarranted assumptions about sex and/or gender, including gender stereotypes. They also maintain that biological research often invokes gendered metaphors that are not recognized *as* metaphors, and that these metaphors carry evaluative content. Feminist philosophers of biology argue that unwarranted assumptions about gender have had a significant negative impact on the directions and content of areas of biological research as well as on social perceptions engendered by them.[2] For example, although feminist and non-feminist philosophers of biology engage reductionism in biological research, feminists place more emphasis on relationships between reductionism and biological determinism—and, in particular, when research and hypotheses propose reductionist and determinist accounts of sex and gender. These kinds of critique notwithstanding, feminists' engagements with biology have been as *constructive* as they have been critical. Feminists have developed what they argue are more empirically warranted research questions, categories and concepts, research methods, causal and explanatory models, and hypotheses, than those they criticize (see e.g., Bleier 1986; Drea and Wallen 2003; Fausto-Sterling 1985/1992; Fehr 2004; Longino 1990; Nelson and Wylie (2004); Schiebinger 1999).

Attention to sex/gender and other socially-salient categories, and in particular to the *political implications* of hypotheses about them, constitutes an important difference at an abstract level in the issues emphasized in the two traditions. Such attention has resulted in differences in the emphasis placed on research that assumes or advocates one or another version of biological or genetic determinism. These differences, in turn, have led to differences in the specific sciences and research programs on which feminist and non-feminist philosophers of biology focus. Feminists have focused more than their non-feminist colleagues on primatology, developmental biology, behavioral and reproductive endocrinology, sexual selection theory, research in neuroscience devoted to finding sex differences in the brain, Human Sociobiology, and Human Evolutionary Psychology. Even when feminist and non-feminist philosophers engage the same issues—for example, cases in which pluralistic explanations might be called for—often practitioners in the traditions understand the philosophical issues at stake differently. So, for example, feminists' analyses

[2]Although the distinction is not without its problems or its critics, I use "sex" to refer to females and males (typically defined by biologists on the basis of their gametes) and "gender" to refer to non-biological characteristics associated with females and males, including women and men: characteristics construed to be "feminine" or "masculine," including behaviors, temperaments, cognitive predispositions, roles, and so forth. Feminists drew and insisted on the distinction early on in the Second Wave of the Women's Movement and it was brought to bear by feminist scientists and science scholars beginning in the 1970s critical of and/or attempting to "disrupt" and/or "replace" the conflation of the two in scientific theorizing in the social and biological sciences.

concerning the appropriateness of pluralistic explanations of various phenomena are more likely to identify and engage relevant sociopolitical as well as empirical issues.

Other differences between the traditions are worth noting. One is that although feminists draw on analyses offered by non-feminist philosophers, non-feminist philosophers of biology often do not draw on or engage feminist analyses of gender and biology. For example, feminists have offered important critiques of hypotheses about sex and gender proposed in Human Sociobiology (HS) and Human Evolutionary Psychology (HEP). Despite their interest in evolutionary biology, non-feminist philosophers often dismiss the significance of feminist critiques of these programs on the grounds that "we all know that" the programs are "unsuccessful"—and, as such, do not warrant critical philosophical attention. (There are important exceptions. They include Buller 2005; Kitcher 1985; Richardson 2007; Vickers and Kitcher 2003.) Yet the issues feminists engage often are, or at least should be, of interest to non-feminist philosophers of biology. Among such issues is the methodological commitment to adaptationism that is a key feature of explanations of sex/gender differences offered in HS and HEP, and that is recognized by many non-feminist philosophers of biology as problematic.

The problems with writing off HS and HEP, and feminist critiques of them, as "unimportant," is that both programs propose hypotheses about sex and gender that carry significant sociopolitical implications (despite protestations to the contrary on the part of their proponents, a point to which I return below).[3] Both are widely-taught at universities and colleges; and their advocates write books and articles addressed to the lay public that receive substantial media attention. In addition, ignoring the claims made in HEP about domestic violence and rape is certainly not commensurate with the relatively new but growing emphasis in both feminist and non-feminist philosophy of science on "socially responsible science," and "socially-relevant" and "socially-responsible" philosophy of science, topics I discuss in my conclusion. In more general terms, a commitment to studying biological science as it is *actually* practiced is incompatible with ignoring well-publicized and influential research programs and claims. Indeed, non-feminist philosophers of biology do not ignore *all* unsuccessful research programs—as evidenced by their engagements with Intelligent Design theory and climate-change deniers.

The more general difference between feminist and non-feminist philosophy of biology these examples illustrate is that, with some important exceptions, there is less interest in and emphasis placed on the social, political, and/or ethical implications of biological research and hypotheses in non-feminist philosophy of biology than there is in feminist philosophy of biology. The exceptions provide examples of how philosophers of biology can address the normative implications of biological research in substantive ways. They include Philip Kitcher's analysis of the social and ethical issues raised by genetic testing (Kitcher 1996); and his sustained critique of HS, which he argues assumes "barroom gender stereotypes" and fails to meet even minimal evidential standards when proposing hypotheses that could have a

[3]I return to this issue in the concluding section.

significant impact on human self-understanding (Kitcher 1985). Jonathan Kaplan engages in analyzes of the empirical problems with, and social and political issues raised by, genetic testing for racial differences—and, in particular, how problematic interpretations of such purported differences are contributing to "race-based medicine" (e.g., Kaplan 2010). And the geneticist Richard Lewontin has offered substantial critiques of hypotheses that explain differences in power and overall wellbeing along the axes of sex, race, and ethnicity, on the basis of biology (e.g., Lewontin 1993). There are other examples; nonetheless, the difference in the amount of philosophical attention the traditions pay to relationships between politics and biology is pronounced.

A third difference, which may be related to those noted, is the increasingly technical nature of non-feminist philosophy of biology. Practitioners often employ mathematical models (to explicate, for example, concepts such as "fitness" and "altruism") and/or engage in analyses of the use of mathematical modeling in biology (e.g., in the utilization of the Hrdy-Weinberg Law, and of game-theoretic approaches to issues in evolutionary theorizing such as altruism—see e.g., articles in Sober 2006). It is not that these emphases are unwarranted or unfruitful; they often result in substantive analyses of biological concepts and models, at times clarifying and even improving them. It is also not the case that feminist philosophers of biology don't make use of technical mathematical models and modeling (e.g., Lloyd 1994). It is to say that the traditions differ in the amount of technical analyses in which they engage and this may affect the philosophical issues they emphasize.

The next three sections focus on themes in feminists' analyses of theoretical, conceptual, and/or practical issues that are more central in the tradition than they are in non-feminist philosophy of biology. The first theme is contextualism: the view that scientists work in specific historical and cultural contexts, and bring experiences that often reflect these contexts (and differences along the axes of socially-salient categories such as gender and race) to bear on their research. Because of this, "who" is theorizing (with "whom" understood non-individualistically), can and often does impact the directions and content of biological research. This, as earlier noted, is recognized by historians and philosophers of biology in terms of historical figures and episodes, but its significance in terms of contemporary biology is much more prominent in feminist philosophy of biology (see e.g., Fehr 2011; Hrdy 1986; Keller 1985; Longino 1990; Nelson 1990). I frame the discussion of feminists' emphasis on contextualism in terms of their identification and critiques of androcentrism in biology; but this is not the only issue that prompts feminists' arguments for contextualism.[4] The second

[4]For example, feminists also cite commitments to uni-directional and "hierarchical" models of causal processes, represented, for example, in Francis Cricks' "Central Dogma" and in models of hormonal effects on behavior in behavioral endocrinology, as an example of how historically-specific views of what constitutes "serious science" impact scientific reasoning. There are stronger versions of contextualism in feminist philosophy of science, including standpoint theory and arguments for the situatedness and partial perspectives of all inquirers (e.g., as advocated, respectively, in Harding 1986 and Haraway 1988).

theme is the presence of gendered metaphors in biological theorizing, and the consequences of these metaphors for the directions and content of the research informed by them. The third theme is the relationships between the biological sciences, and political and ethical issues and concerns.

Each section includes examples of research feminists have engaged and arguments they have offered about the implications of their analyses. The examples are accessible and representative of analyses that led to some common lines of argument in feminist theorizing about biology and other sciences. However, neither the examples, nor the lines of argument summarized, capture all the areas of biological research, or all the conceptual and empirical issues, that feminist philosophers of biology engage. The aim is to provide a useful introduction into the evolving and nuanced nature of feminist engagements with biology, and into ways in which these engagements differ from non-feminist engagements.

In my conclusion, I identify ways in which feminist and non-feminist philosophy of biology might fruitfully engage each other. As earlier noted, I emphasize three recent themes of growing interest in philosophy of science: how to encourage "socially responsible science," and "socially relevant" and "socially responsible philosophy of science." Significant overlaps in the arguments so far put forward in support of these projects, as well as overlaps with issues feminists have long emphasized, can and should encourage collaborations between practitioners of feminist and non-feminist philosophy of biology—something that is already taking place among feminist and non-feminist philosophers of science.

4.2 Contextualism: How *Who* Is Theorizing Can Matter

Although early feminist critiques of biology often focused on sexist assumptions and hypotheses—that one sex (virtually always males) is superior—by the 1970s feminists began to recognize and analyze a more subtle way in which sex/gender is relevant to biology: they identified the role and consequences of androcentrism in research focuses, questions, categories, methods, and hypotheses. Androcentric or "male centered" approaches focus on behaviors, characteristics, and spheres of activity associated with males or men, and ignore or downplay those associated with females or women. Feminists identified androcentrism and problems resulting from it in a number of biological research programs, and used these findings to argue that the historical and social contexts within which science is undertaken—particularly when, as long was the case, science communities did not encompass diverse perspectives and experiences—can help shape the questions addressed, categories assumed, methods employed, interpretations of data, and hypotheses proposed. We begin with some examples of androcentrism feminists identified.

In developmental biology, feminists pointed to a lack of investigation, continuing well into the 1980s, into the processes that result in females, in models purportedly describing *human* fetal development (see e.g. Bleier 1984; Fausto-Sterling 1985/1992; Gilbert and Rader 2001). The models, which enjoyed broad acceptance, emphasized the role of the Y chromosome in initiating the masculinization of an embryo by causing an influx of androgens that, together with Mullerian-inhibiting substances, led to the development of male internal gonads, and the role of the latter in producing hormones that lead to the development of male external genitalia.[5] The study of how hormones more common in females (primarily, estrogens) are or might be involved in the development of the ovaries and uterus, and eventually female external genitalia, was largely absent. Perhaps developmental biologists assumed that female development was "the default trajectory." But, feminists argued, this assumption does not warrant neglecting the processes involved in female fetal development.

In behavioral endocrinology, again well into the 1980s, the role of pre-natal and pubescent male hormones in organizing the developing male brain and leading to "masculine" behaviors and temperament, were a common focus. As in developmental biology, little if any research was devoted to the role of estrogens in the organization of the developing fetal brain (female or male), or into the role of estrogen in predisposing or determining what were assumed to be "sex- or gender-specific" behaviors. Rather, the effects of too much by way of male hormones was proposed to explain what were assumed to be "abnormal" female behavior. For example, one hypothesis proposed that "tomboyism"—a decidedly ill-defined and, in several respects, problematic "scientific" category—is the result of an over-abundance of androgens in the prenatal development of girls described as exhibiting "it." Another proposed that female homosexuality is a result of over exposure to circulating androgens during fetal development. Feminists criticized the androcentrism of this research, the stereotypes involving gender and sexuality its hypotheses assumed, and the lack of attention to interactions between hormone levels and behavior—interactions that effectively undermine the uni-directional causal models and biological determinist explanations of male (and in some cases, female) behavior (see e.g., Bleier 1984; Longino 1990).

In neuro endocrinology, prenatal and pubescent androgens were often take to explain purported sex-differences in the cognitive abilities and roles of women and men—even though much of the research on which such hypotheses were based on research into developing rat fetal brains and behaviors exhibited by caged rodents (Bleier 1984; Fausto-Sterling 1985/1992). Focusing on the role of prenatal "male" hormones, researchers sought answers to questions such as "Why are there so many more men than women in math and science?" (Bleier 1984), ignoring the formal and informal barriers that, feminists argued, were far more likely explanations of the disparities.

Androcentrism in primatology was another focus of feminist philosophy of biology. In his initial arguments for HS, Wilson (1975) proposed that the behaviors of closely-related non-human primate could provide insights into early and current

[5]One focus of feminist critiques of the models under discussion, and of research in reproductive and neuro endocrinology next discussed, is the assumption of sexual dimorphism. This is beyond the scope of the current discussion, but see Fausto-Sterling (1985/1992).

human behavior. Wilson and other human sociobiologists were particularly interested in what primate behavior could tell us about sex-differentiated behaviors and mating strategies, which they assumed would result from sexual selection and be common across species. Feminists argued that the primatology on which sociobiologists drew was itself deeply androcentric: early and mid-twentieth century primatologists focused on the behaviors of and relationships between males as they assumed them were the most consequential to the social dynamics of primate groups. Field and laboratory research emphasized male-dominance hierarchies and "alpha males," and paid relatively little attention to female primates. In addition, feminists argued that when researchers did describe female behavior, the descriptions (like those of male primates) were often couched in gender-stereotypical terms. Females were "observed" to be more passive, dominated by males, and largely involved in reproductive activities that were not viewed as consequential for social dynamics. Finally, some feminist primatologists pointed out that the primate species most focused on for insights into early and current human behavior, the ordinary chimpanzee, reflected "selection" on the part of primatologists. A second species of chimpanzee, Bonobos, as closely related to humans as feminists argue, the role of female hierarchies and the prevalence of cooperative, rather than aggressive, behaviors, might have led to different hypotheses about human behavioral capacities and the recognition of significant variation in the social dynamics characterizing primate species (Fehr 2011).

These are just a few examples of the androcentrism in the biological sciences that feminists identified and critically engaged. Beyond the obvious emphasis on biological assumptions about the sexes and/or genders, in what ways do these and similar examples indicate differences between feminist and non-feminist philosophy of science? Let's return to the theme that is the focus of this section: namely, feminist arguments for contextualism and that "who" is theorizing can make a difference to the directions and content of biological research. Three arguments were common. In one, feminists attributed androcentrism to the prevalence, before the lifting of formal and informal barriers to women, of men in the sciences. In a related argument, feminists cited social beliefs and values about sex/gender that characterized the larger social and cultural contexts in which science is undertaken, as contributing to androcentrism. More controversially in terms of traditional views of science, in a third line of argument, feminists attribute the influx of women (and men) whose political beliefs and values were in part shaped by the Women's Movement, as enabling the recognition of androcentrism and the alternative questions, methods, and hypotheses that feminist scientists proposed.

Although as earlier noted, non-feminist philosophers who study the history of biology, as we saw in the case of Darwin scholarship, often identify relationships between aspects of biological theorizing and then current social and political, as well as scientific, contexts, far less attention is paid to relationships between recent or current research questions and emphases, on the one hand, and sociopolitical beliefs and values, on the other (Kitcher 1985; Ruse 2013 are important exceptions). And for many scientists and philosophers of science, the social identity—gender, race, culture, ethnicity, and the like—of individual scientists has been assumed to be irrelevant to the questions they pursue, categories they assume, methods they employ, and hypotheses they propose—at least when things are going as they should. Many believed (and at least some philosophers, scientists, and lay people continue to believe) that scientific methods, shared epistemic values taken to be "constitutive" of science, peer-review mechanisms, and other features of science's social organization, will filter out any "bias" or unwarranted assumption an individual scientist or group might unwittingly bring to bear on their engagements with nature.[6] Such inter-subjectivity, together with the longstanding and related assumption that science is "value-free" (with the exception of cognitive or epistemic values), would seem to rule out the possibility that who is theorizing *could* matter—at least in the case of "good" science. But the feminist research we have considered indicates otherwise. Good science can and often has reflected aspects of the specific social and historical context within which it is undertaken, including the "social identities" of scientists engaging in it.

To critics of feminists' arguments linking social context to scientific research (e.g., Haack 1993; Pinnick et al. 2003), neither the recognition of androcentrism and the problems to which it led, nor the proposal of alternative (and, arguably, more warranted) research questions and hypotheses, has anything to do with *feminism* and/or with "who" scientists are. Three arguments for this conclusion were common. The least plausible was that it was a *coincidence* that androcentrism was recognized when women entered science in larger numbers and brought their experiences of the Women's Movement to bear on the hypotheses, assumptions, and methods of their fields. Another took the recognition of androcentrism as just an example of how science is inherently "self-correcting" to undermine claims that there were relationships between feminism and the recognition of androcentrism. A third line of argument maintained that the problems feminists identified only involved cases of "bad science;" as such, their critiques were not relevant to "science as usual."

I return to questions about relationships between politics and science raised by the third line of argument in Sect. 4.4. Here I consider whether coincidence or "self-correction" is a more likely explanation than that "who" is theorizing can, and in the cases at hand and others of which they are representative, do, matter. Fortunately, we don't need to decide between these competing explanations in the absence of evidence. In *Primate Visions* (1989), primatologist Donna Haraway cites interviews with, and the writings of, four primatologists who maintain that feminism had a significant impact on their research focuses, observations, and hypotheses. I focus on one of the primatologists, Jeanne Altmann, who made significant contributions to methodologies used in field and laboratory studies in primatology. In part due to problems she identified in methods employed by field primatologists, she was able to observe that female baboons control and manipulate

---

[6]I emphasize unrecognized and unwitting bias because cases in which scientists (or others) deliberately manipulate research questions, hypotheses, or tests on the basis of bias, are, from an epistemological point of view, uninteresting.

key features of the social dynamics of their troops. I then briefly discuss parallels between her research focus and the relationship she draws between it and feminism, with the research and arguments offered by primatologist Sara Blaffer Hrdy.

According to Altmann, her study of *female* primates—in particular, the relationships between, female savannah baboons and their offspring—was something she initially wanted to avoid. She worried that the research emphasis would be perceived as reflecting a non-objective "empathy" with her research subjects. (Interestingly, the emphasis on male behavior and relationships then emphasized in primatology was *not* viewed as reflecting "empathy" with their subjects on the part of male primatologists.) She also worried that such research would be seen as unimportant given the prevalence of male-centered analyses in twentieth century primatology. Altmann also notes that, at the time (and of course some scientists and philosophers still hold the view), for a scientist to acknowledge that her research was in part related to her *feminism*, was also viewed as deeply problematic.

Yet, Altmann argues, eventually her own complex self-identity—as a scientist, feminist, and mother—did lead to her focus on female baboons. "Increasingly," Altmann reports, "it was screaming at me. These are the most interesting individuals; [interactions between baboon mothers and offspring] have the most evolutionary impact; this is where the ecological pressures are" (Haraway 1989: 312). Altmann's field studies revealed that baboon mothers engage in what Altmann describes as "juggling" between competing priorities similar to those that "dual-career" women, women who like herself are mothers and pursue careers, engage in. Using the metaphor of "budgeting"—and being quite specific that this *is* a metaphor, an issue we focus on in the next section—Altmann's observations revealed that females' budgeting and interactions with both female and male members of their troop, had important consequences for social dynamics.[7] Of budgeting she said "You could call it budgets or something else […] The issue is hierarchy of demands and the immediate consequences of these demands. What is and is not flexible in one's life is to me terribly fascinating, terribly important biologically, and also important for my experience as a human being, a woman, a mother" (Haraway 1989: 314). Further breaking from the emphasis on male-centered analyses of baboon social dynamics, Altmann proposed that the "high drama" characterizing encounters and relationships between male baboons, the then primary focus of primatologists, is not nearly as important to the social dynamics of baboon troops as the "micro-practices" of females.

Altmann's accounts of the relationships between her feminism and her study of female primates are representative in their general outlines of the other primatologists on which Haraway focuses. Primatologist Sara Blaffer Hrdy studies female langurs and observed the various techniques they use to manipulate their male counterparts. Sometimes they mate with a number of males so that paternity is

[7]As discussed in the next section, metaphors are abundant in the biological sciences—and some argue in science generally. One major problem arises when scientists don't recognize the metaphors they invoke *as metaphors*. Altmann was specific that her notion of "budgeting" *is* a metaphor.

unclear and several males end up protecting and providing for their offspring; and they can conceal estrus or bring about cyclical estrus to promote their reproductive success. Using Parental Investment Theory, a theory we will see feminists are generally very critical of, Hrdy turned the tables on traditional accounts of how it leads to "coy" females given their greater investment in offspring. Hrdy argues that her field studies of langurs indicate that relatively larger female investment can lead to polyandry and other behaviors to manipulate males into investing in offspring. Like Altmann, Hrdy attributes the significant differences in accounts of female primate behavior that came to characterize primatology beginning in the 1970s, to the influx of women into the field, many of whom identified themselves as feminists (Hrdy 1986).

Although the details differ in relation to the area of biology, Altmann's and Hrdy's arguments are representative of accounts other feminist biologists provide of how they recognized problems involving assumptions about sex/gender in their fields, engaged in detailed research that challenged them, and proposed alternative hypotheses (e.g., Bleier 1984; Fausto-Sterling 1985/1992; Keller 1985). Contrary to the "coincidence" and "self-correction" hypotheses, in the view of many field and laboratory biologists, feminism has had a significant impact on the directions of their research, their observations, and their hypotheses. And contrary to arguments that feminism, or a political commitment of any kind, necessarily compromises a scientist's research, the critiques and alternatives discussed in this section make it clear that *evidence* is the issue most emphasized by these scientists both in terms of the critiques of traditional approaches and of the alternative methods and hypotheses they propose.

4.3 The Presence and Role of Gendered and "Evaluatively-Thick" Metaphors

Gendered metaphors abound in the biological sciences. They include, but are by no means exhausted by, descriptions of the males of many species as "promiscuous," "undiscriminating," and "active;" and the females as "coy," "discriminating," and "passive." As they have functioned in evolutionary biology, these particular metaphors have roots in Darwin's accounts of sexual selection in *The Origin* and *The Descent*, although the associations of males with activity and females with passivity dates at least to Aristotle. In the 1960s, another metaphor—that of "parental investment"—was the core of a theory claiming to explain the sex-differentiated mating strategies Darwin proposed (male competition for females and female choice). The metaphor combines economic and biological concepts as does the theory to which it led, Parental Investment Theory (PIT), which would become a mainstay of HS and HEP. According to PIT, gametic dimorphism (the difference in terms of size and number of male and female gametes), and additional features of female investment in offspring (including egg sitting, gestation, and lactation in mammals), explain the two components of Darwin's account of sexual selection. In keeping with the economic metaphor, sperm are described as "cheap," eggs as relatively "expensive," and females as a "scarce resource." These economically-infused "biological" differences are, in turn, taken to explain why males are "promiscuous" (as their genes can be replicated by mating with many females, and for whom the issue of investing time and energy to offspring that might not be their own is important) and why females are "discriminating" (who, it is assumed, not only seek to mate with a fit male given the limits on how many offspring they can produce; but are also dependent on males to provide for and protect them and their offspring).

We saw earlier that Sara Bladder Hrdy uses parental investment theory to arrive at very different conclusions about how females, given their higher investment in offspring, behave from those predicted by PIT, observed by mid-twentieth century primatologists, and assumed in HS and HEP. In the langurs troops Hrdy observed, the heavier investment incurred by females does not lead to coy behavior, but manipulative behavior that has a significant impact on social dynamics. We should also note that sexual selection, both as Darwin formulated the hypothesis and advocates of PIT build from it, is now controversial in many quarters, although neither human sociobiologists nor human evolutionary psychologists acknowledge this. Feminists and non-feminists have argued that sexual selection is either not an evolutionary mechanism at all or, at most, a relatively insignificant mechanism (see e.g., Spencer and Masters 1992). In many critiques of sexual selection theory, morphological characteristics and behaviors previously explained on the basis of sexual selection are argued to be the result of natural selection. So, too, the metaphor of "cheap sperm" has been criticized on the grounds that sperm production is itself "costly" (see e.g., Spencer and Masters 1992). We turn to issues involving the political implications of explanations of sex-differentiated mating strategies as they figure in HEP in section four. Here, I draw attention to the metaphors Darwin's account of sexual selection and PIT invoke.

The metaphors of "active males" and "passive females" have had consequences for theorizing in other areas of biology. Martin (1991) and the Biology and Gender Study Group, led by embryologist Scott Gilbert (Biology and Gender Study Group 1989), document how textbook accounts of fertilization, used for decades in medical schools as well as college courses in cell biology, invoke these gendered metaphors. As Martin shows, many such accounts strongly parallel the plot and actors in the fairy tale, "Sleeping Beauty." Sperm are portrayed as the actors of the process/story (akin to the knights seeking to kiss and awaken the sleeping princess), while a passive egg (akin to the sleeping princess) awaits the winner of the "competition" between sperm. As Martin and the Biology and Gender Study Group argue, these accounts do not describe how an egg repels some sperm and capacitates others. The Biology and Gender Study Group associates the eventual recognition of the egg's active and important role in fertilization with changing views about sex/gender, and maintain that in this case and others, feminist critiques make significant contributions to cell biology (see Gilbert and Rader 2001).

It is a common view in philosophy of science, including philosophy of biology, that metaphors play a substantive role in scientific theorizing—a role that, so long as metaphors are recognized *as metaphors*, is not only unproblematic but can advance theorizing. Feminists argue that the gendered metaphors invoked in the biological sciences are problematic and compromise, rather than advance, theorizing. For one thing, most often the gendered metaphors invoked in biology are not recognized *as metaphors* but are taken to be literal empirical descriptions. Feminists also argue that gendered metaphors often reflect and reinforce unwarranted gender stereotypes and contribute to androcentric hypotheses and biological determinist accounts of male and female behavior and capacities.

Feminist philosopher Elizabeth Anderson's notion of "evaluatively-thick concepts" is helpful in understanding the concerns feminist express about the role and consequences of gendered metaphors. As Anderson (2004) defines an "evaluatively-thick concept," it is one that includes *both* empirical and evaluative content.[8] Like the gendered metaphors we are considering, Anderson argues that often a concept is not recognized as evaluatively-thick by scientists who employ it. To illustrate the nature and role of evaluatively-thick concepts, Anderson uses a case study that compares non-feminist approaches to divorce with the approach taken by researchers who identify themselves as feminists. She argues that the evaluatively-thick concepts that informed traditional approaches to divorce led to different observations and findings from those resulting from the research feminists undertook.

Many traditional researchers studying divorce take the financial effects on women to be "harmful," and there is empirical evidence to support the assumption. Given differences in men's and women's incomes, and that men are more likely to engage in work outside the home, in general, divorce has a more negative impact on women's financial status than it does on men. In traditional research, this phenomenon is categorized as straightforwardly "harmful." But, Anderson notes that when scientists brought specifically-feminist perspectives to their analyses of divorce—and, in particular, *an agnostic perspective* on the effects of divorce on women and children—they asked questions of the women they interviewed that were not designed to elicit information only about the negative effects of divorce. As a result, they learned that women often felt that their loss in income, although actual, was counter-balanced by their new independence and, in some cases, new opportunities to work outside the home. Of course, the agnostic attitude towards divorce that the feminist researchers took is also evaluative; the important difference, as Anderson makes clear, between their approach and that of traditionally-informed research, is that this evaluative component was clearly recognized by the feminist researchers.

Anderson cites other examples from the case study, including how the traditional hypothesis that the effects of divorce on children are harmful because their parents no longer live together, was challenged when children were asked about that change in ways that did not invite them to talk about how its negative impact. Many

[8]Anderson is not the first to argue for "evaluatively-thick concepts." She is the first to find such concepts at work in science and to provide a feminist analysis of their role and consequences.

children expressed relief at no longer being in tense environments and/or witnessing fights between their parents. Anderson uses the case study to argue that a concept such as "a harmful effect" carries both empirical and evaluative content. Not recognizing this and taking the concept of "harmful effects" as purely descriptive, resulted in traditional researchers not pursuing whether there were positive as well as negative effects of divorce-related phenomena. As Anderson argues, the juxtaposition of traditional hypotheses with those generated by feminist researchers who brought an agnostic attitude towards divorce, illustrates how "evaluatively-thick" concepts can yield conclusions that are themselves evaluative as well as empirical.

I suggest that feminists' critiques of the role and consequences of gendered metaphors in biology often seek to demonstrate that the metaphors carry both empirical and evaluative content. We have noted the common associations of males with activity and aggression, and of females with passivity and dependence. Each quality has evaluative as well as empirical connotations, and at least in terms of contrasts between "active" and "passive," and between "independent" and "dependent," the first of each pair has been taken to be of higher value. We also considered the metaphors of parental investment, active and promiscuous males, and coy and passive females, as they have figured in historical evolutionary theorizing and in the more recent programs of HS and HEP. And in our discussion of how "who" is theorizing can matter, we learned how these metaphors were long taken to be key empirical features of primate social dynamics, and of how research and observations of feminist primatologists challenged the accounts of behavior and social dynamics in which the metaphors were invoked.

We have considered a few examples of gendered metaphors in biology that feminists have identified as metaphors, rather than literal descriptions, of animal and human behavior and psychology. I have argued that recognizing that these and other gendered metaphors are evaluatively thick—that they carry both evaluative and empirical content—is useful in understanding feminists' concerns about and critiques of them. In the next section, I will argue that identifying concepts, metaphors, and hypotheses about sex/gender that are evaluatively-thick, and not simply or primarily empirical, demonstrates that, contrary to recent claims by advocates of Human Evolutionary Psychology (e.g., Buss 2004), feminists who cite political and ethical implications of hypotheses advanced in the program, are not guilty of committing the naturalistic fallacy. The hypotheses feminists criticize are not, as their proponents claim, "purely empirical statements about how things are." They carry significant evaluative content.

4.4 Biology and Politics

The issues feminists engage have raised the question of whether the biological research and hypotheses feminists criticize is appropriately described as "bad science." We have considered two arguments that seek to deny any relationship between *feminism* and the problems feminists identified: the "coincidence" argument and the argument that science is inherently "self-correcting." A third argument is that the cases on which feminists focus are straightforwardly "bad" or "unsuccessful" science, and thus provide little if any insights into good science or science "as usual." In one or more ways, the sciences feminists criticize fail to meet the norms of good science.

Responses to such arguments are obvious. The research and hypotheses that feminists identified as androcentric and/or informed by gendered metaphors were, in all other respects, good science. They reflected discipline-specific research questions and accepted methods, and yielded results that, even if incomplete or in some cases mistaken, were compatible with what was *recognized as available evidence*—just as contextualism would predict. In addition, there is no evidence that scientists who brought untested background assumptions about sex/gender to bear on their methods, and hypotheses, were aware that these assumptions were unwarranted. These were not recognized as problematic *until* women and men sensitive to gender issues began entering biology. Prior to these changes in the demographics of science communities, there was little if anything to suggest to a relatively monolithic population in terms of their experiences and assumptions, that their assumptions about sex/gender were possibly flawed and, in any event, warranted scrutiny.

Finally, it is simply unreasonable to write off a large amount of twentieth-century biology as "bad science." A far more likely explanation is that the research in question reflects a range of social, cultural, and scientific contexts. If as feminists and others argue, the role of such contexts in helping to shape scientific practice is unavoidable, then concluding that any research or inquiry that reflects them constitutes "bad science," is inappropriate.

I conclude this section on the theme of biology and politics with a brief discussion of HEP. It is an interesting case for several reasons. First, in contrast to the research feminists have criticized that cannot be written off as "bad science," the conceptual and methodological problems that characterize this research program are recognized by many biologists and philosophers of biology—even though, as I earlier noted, few have taken up specifically feminist critiques of it. Critics have faulted its methodology as adaptationist, often circular, and as not making use of accepted "historical methods" to identify human "cognitive adaptations" that they claim emerged in the Pleistocene. On the conceptual level, its commitment to cognitive modules (and to there being many of them) to solve adaptive problems in the Pleistocene has been criticized by cognitive scientists and philosophers of mind; and its commitments to the existence of "universal" cognitive predispositions, and based on such "universality" to their "innateness," have also been criticized. It is, I and others have argued, deeply flawed (see e.g., Buller 2005; Nelson 2003; Richardson 2007; Vickers and Kitcher 2003). Yet this, as I earlier argued, is *not* reason to ignore this research program. Its hypotheses about sex/gender—including, but not limited to sex-differentiated mating strategies and dispositions, domestic violence, and rape—carry significant sociopolitical implications (e.g., Thornhill and Palmer 2000; Wilson and Daly 1998). I have elsewhere argued that, given its high profile, and the cognitive authority granted to and exercised by scientists, evolutionary theorists, cognitive scientists, and philosophers of biology need to critically engage the program more than they have to date (Nelson 2003).

Another reason why HEP is interesting is that, recently, in response to critiques by feminists, its advocates have launched arguments that their feminist critics are guilty of committing the naturalistic fallacy. HEP theorists, it is argued, are just reporting "how things *are*," not how they *should be*. Feminists, they continue (apparently assuming there aren't feminist *scientists*), are concerned with how things *should be* and wrongly assume that by telling us how things *are*, advocates of HEP are telling us how they should be. It is true that in their explanations concerning sex/gender, human evolutionary psychologists often add the caveat "we are not claiming that the cognitive predispositions selected for in the Pleistocene are currently adaptive." But this does little if anything to undermine or ameliorate the force that claims about "adaptations" and "innate predispositions" carry. Nor are human evolutionary psychologists in a position to control how members of the lay public will understand them, or qualified to engage the social and ethical issues raised by their hypotheses.

As importantly, in providing explanations, based on PIT, of sex-differentiated mating strategies and dispositions, the strategies and dispositions proposed are as evaluatively-thick as the theory on which they are based. So explanations invoking them are not just literal descriptions, despite what human evolutionary psychologists claim. Domestic violence and rape are no less evaluatively-thick. Indeed, arguably, the negative evaluative content most associate with domestic violence and rape is diminished by explanations that they are the product of natural selection in the Pleistocene and inherited by modern humans.

We have seen that relationships between biology and politics are a key focus of feminist philosophy of biology. As noted in the introduction, interest in the political implications of biological research, models, and hypotheses has been less common, though not totally absent, in non-feminist philosophy of biology. The interesting philosophical questions raised by the relationships that feminist engage include how to contribute to more nuanced views of their practices among scientists, and more nuanced views of science among the lay public and policy-makers. As I address in the next section, there is recent and growing interest in these issues among feminist and non-feminist philosophers of science.

4.5 Prospects for Future Engagements: Socially-Responsible Science, and Socially-Relevant and Socially-Responsible Philosophy of Science

There is an obvious sense in which the issue of "socially-responsible science" has been a prominent theme in feminist philosophy of biology. In their critiques of, and proposed alternatives to, androcentrism, biological determinism, gender stereotypes and gendered metaphors, many feminists emphasize the problematic nature of the evidence supporting hypotheses about sex/gender and the role of the hypotheses in shaping public perceptions. Janet Kourany documents a number of these arguments in *Philosophy of Science after Feminism* (2010) in which she argues that philosophers of science should support and help to motivate "socially-responsible science" (SRS). On Kourany's account, SRS is science that is focused to an appropriate degree on improving human wellbeing. For philosophers of science, emphasizing a need for sciences of this sort means breaking with the tradition's mid-twentieth century view that science is apolitical and value free, and with its emphasis on issues internal to science. And although many who accepted these views, as well as many who rejected them, attributed them to logical empiricism, recent research demonstrates that there were divisions among participants in that research program concerning the role of science and of philosophy of science in promoting justice and human well-being (e.g., Okruhlik 2004).

By the mid-1990s and continuing to the present, interest in and attention to issues involving "socially-relevant philosophy of science," and "socially-responsible philosophy of science" have grown in both feminist and non-feminist philosophy of science. The interest and attention are attested to by conferences, colloquia, special journal issues, and collections of essays. In 2008 a mini-conference held concurrently with the Pacific APA meetings, titled "Making Philosophy of Science More Socially Relevant," was organized by feminist philosophers Nancy Cartwright and Helen E. Longino. Participants were asked to address social, political, and ethical issues identified in philosophical analyses of areas of scientific research. Also in 2008, the Philosophy of Science Association hosted the workshop, "How Philosophers of Science Can Take Up Socially Relevant Roles," which identified some aspects of philosophy of science that would make it difficult to in fact engage in such roles. The conference and workshop led to a special issue of *Synthese* devoted to Socially-Relevant Philosophy of Science (or SRPOS as it came to be called) edited by Fehr and Plaisance (2010). Contributors addressed topics of scientific research and scientific practices that, in the editors' words, "are *directly relevant to public welfare*, such as investigations of race and genomics, biomedical research, or special interest science" (Fehr and Plaisance 2010: 302). They also addressed issues raised by the existence of "various *stakeholder* groups," including policy-makers, members of marginalized groups, and "various publics;" and which practices and venues would enable philosophers of science interested in undertaking SRPOS to have the most social impact (Fehr and Plaisance 2010: 303). Many, but certainly not all, of those who attended the 2008 conference and 2008 workshop, or published an article in the special *Synthese* issue, identify themselves as feminists. Indeed, the research topics and practices both feminist and non-feminist contributors addressed went well beyond sex and gender.

In 2012, the conference "The Social Relevance of Philosophy of Science" was organized by philosophers of science Martin Carrier and Don Howard and hosted by the Center for Interdisciplinary Research at Bielefeld University. Participants engaged issues involving the relationships between science and society, including how to analyze and make public the evidence (or lack thereof) for theories that proposed biological or genetic determinism; how to conceptualize the "moral" responsibility that come with engaging in science; and making information available to the lay public about how research in an emerging science (e.g., nano science and climate science) can yield competing models and predictions, highlight some questions and "eclipse" others, and evaluate the epistemic status of heuristic strategies in such fields.

The growing interest in philosophy of science to engage issues that are simultaneously epistemic and ethical, including how to practice and/or motivate socially responsible science and socially responsible philosophy of science, is not limited to feminist philosophers of science. Because of this, and given the parallels between the issues recognized as significant and pressing in these projects, and issues feminist philosophers of biology have emphasized, there is reason to hope for more substantive engagements between the feminist and non-feminist traditions in philosophy of biology. After all, the biological sciences are certainly among the sciences that have had and will continue to have a significant impact on human wellbeing—sometimes positively, and regrettably sometimes not.

References

Anderson, E. (2004). Uses of value judgments in science: A general argument, with lessons from a case study of feminist research on divorce. *Hypatia, 19*(1), 1–24.

Biology and Gender Study Group. (1989). The importance of feminist critique for contemporary cell biology. In N. Tuana (Ed.), *Feminism and science* (pp. 172–187). Bloomington: Indiana University Press.

Bleier, R. (1984). *Science and gender: A critique of biology and its theories on women*. Elmsford, NY: Pergamon Press.

Bleier, R. (Ed.). (1986). *Feminist approaches to science*. New York: Pergamon Press.

Buller, D. (2005). *Adapting minds: Evolutionary psychology and the persistent quest for human nature*. Cambridge, MA: MIT Press.

Buss, D. (2004). *Evolutionary psychology: The new science of the mind*. Boston: Allyn and Bacon.

Darwin, C. (1871). *The descent of man and selection in relation to sex*. London: John Murray.

Drea, C., & Wallen, K. (2003). Female sexuality and the myth of male control. In C. Travis (Ed.), *Evolution, gender, and rape*. Cambridge, MA: MIT Press.

Fausto-Sterling, A. (1985/1992). *Myths of gender: Biological theories about women and men*. New York: Basic Books.

Fehr, C. (2004). Feminism and science: Mechanism without reductionism. *National Women's Studies Association Journal, 16*, 136–156.

Fehr, C. (2011). Feminist philosophy of biology. In E. N. Zalta (Ed.), *The stanford encyclopedia of philosophy*. http://plato.stanford.edu/archives/fall2011/entries/feminist-philosophy-biology. Accessed Jan 5, 2015.

Fehr, C., & Plaisance, K. (2010). Making philosophy of science more socially relevant: An introduction. *Synthese, 177*(3), 301–316.

Gilbert, S., & Rader, K. (2001). Revisiting women, gender, and feminism in developmental biology. In A. Creager, E. Lunbeck, & L. Schiebinger (Eds.), *Feminism in twentieth-century science, technology and medicine* (pp. 73–97). Chicago: University of Chicago Press.

Haack, S. (1993). Epistemological reflections of an old feminist. *Reason Papers, 18*, 31–43.

Haraway, D. (1988). Situated knowledges: The science question in feminism and the privilege of partial perspective. *Feminist Studies, 14*, 575–599.

Haraway, D. (1989). *Primate visions: Gender, race, and nature in the world of modern science*. New York: Routledge.

Harding, S. (1986). *The science question in feminism*. Ithaca: Cornell University Press.

Hrdy, S. B. (1986). Empathy, polyandry, and the myth of the coy female. In R. Bleier (Ed.), *Feminist approaches to science* (pp. 119–146). New York: Pergamon Press.

Hubbard, R. (1983). Have only men evolved? In S. Harding & M. Hintikka (Eds.), *Discovering reality: Feminist perspectives on epistemology, metaphysics, methodology, and philosophy of science* (pp. 45–70). Dordrecht: Reidel.

Kaplan, J. (2010). When socially determined categories make biological realities: Understanding black/white health disparities. *Monist, 93*(2), 283–299.

Keller, E. F. (1985). *Reflections on gender and science*. New Haven: Yale University Press.

Kitcher, P. (1984). 1953 and all that: A tale of two sciences. *The Philosophical Review, 93*(3), 335–373.

Kitcher, P. (1985). *Vaulting ambition: Sociobiology and the quest for human nature*. Cambridge, MA: MIT Press.

Kitcher, P. (1996). *The lives to come: The genetic revolution and human possibilities*. New York: Simon and Schuster.

Lewontin, R. C. (1993). *Biology as ideology: The doctrine of DNA*. New York: Harper Perennial.

Lloyd, E. A. (1994). *The structure and confirmation of evolutionary theory*. Princeton: Princeton University Press.

Lloyd, E. A. (2005). Why the gene will not return. *Philosophy of Science, 72*, 287–310.

Longino, H. E. (1990). *Science as social knowledge: Values and objectivity in scientific inquiry*. Princeton: Princeton University Press.

Martin, E. (1991). The egg and the sperm: How science has constructed a romance based on stereotypical male-female roles. *Signs, 16*, 485–501.

Nelson, L. H. (1990). *Who knows: From Quine to feminist empiricism*. Philadelphia: Temple University Press.

Nelson, L. H. (2003). The descent of evolutionary explanations: Darwinian vestiges in the social sciences. In S. P. Turner & P. A. Roth (Eds.), *The Blackwell guide to the philosophy of the special sciences* (pp. 258–290). London: Blackwell.

Nelson, L. H. & Wylie, A. (2004). Special issue of hypatia on feminism and science. *Hypatia, 19*(1).

Okruhlik, K. (2004). Logical empiricism, feminism, and Neurath's auxiliary motive. *Hypatia, 19* (1), 48–72.

Pinnick, C., Koertge, N., & Almeder, R. (Eds.). (2003). *Scrutinizing feminist epistemology: An examination of gender in science*. New Brunswick, NJ: Rutgers University Press.

Richardson, R. (2007). *Evolutionary psychology as maladapted psychology*. Cambridge, MA: MIT Press.

Ruse, M. (2008). *Darwinism and its discontents*. Cambridge: Cambridge University Press.

Ruse, M. (2013). *The Gaia hypothesis: Science on a pagan planet*. Chicago: University of Chicago Press.

Schiebinger, L. (1999). *Has feminism changed science?*. Cambridge, MA: Harvard University Press.

Sober, E. (Ed.). (2006). *Conceptual issues in evolutionary biology* (3rd ed.). Cambridge, MA: MIT Press.

Sober, E., & Wilson, D. S. (1999). *Unto others: The evolution and psychology of unselfish behavior*. Cambridge, MA: Harvard University Press.

Spencer, H. G., & Masters, J. C. (1992). Sexual selection: Contemporary debates. In E. F. Keller & E. A. Lloyd (Eds.), *Keywords in evolutionary biology* (pp. 294–301). Cambridge, MA: Harvard University Press.

Thornhill, R., & Palmer, C. (2000). *A natural history of rape*. Cambridge, MA: MIT Press.

Vickers, A. L., & Kitcher, P. (2003). Popsociobiology reborn: The evolutionary psychology of sex and violence. In C. Travis (Ed.), *Evolution, gender, and rape* (pp. 139–168). Cambridge, MA: MIT Press.
Wilson, E. O. (1975). *Sociobiology: The new synthesis*. Boston: Harvard University Press.
Wilson, M., & Daly, M. (1998). Lethal and nonlethal violence against wives and the evolutionary psychology of male sexual proprietariness. In R. E. Dobash & R. P. Dobash (Eds.), *Violence against women: International and cross-disciplinary perspectives* (pp. 199–230). Thousand Oaks, CA: Sage.

Chapter 5
Feminist Values, Commercial Values, and the Bias Paradox in Biomedical Research

Kristen Intemann and Inmaculada de Melo-Martín

5.1 Introduction

Feminist approaches to philosophy of science have typically involved two claims that some take to be in tension with one another. First, feminists have argued that sexist, racist, classist, and heteronormative values and assumptions are problematic in scientific reasoning because the influence of those values often leads to epistemically bad theories (Fausto-Sterling 1992; Harding 1986; Keller 1985; Longino 1990). On the other hand, feminist theorists have often proposed approaches to science that advocate the incorporation of feminist values in directing scientific reasoning and practices (de Melo-Martin and Intemann 2011; Harding 2008; Kourany 2010). Thus this seems to present feminist approaches with a "bias paradox:" if sexist values are bad because they are *partial*, then how can feminists claim that we ought to be partial towards feminist values? (Antony 1993).

This paradox appears to be particularly pressing in the current context of biomedical research, where commercial interests and values pose significant concerns about bias. For instance, a variety of studies have found that research funded by industry is significantly more likely to report positive outcomes than studies funded by not-for-profit organizations (Bourgeois et al. 2010; Khan et al. 2008; Lundh et al. 2012; Nkansah et al. 2009; Sismondo 2008). Similarly, there is concern that profit-driven interests may lead research in directions that are unresponsive to basic public health needs or that neglect the health needs of marginalized groups (De Winter 2012; Pogge 2009; Reiss and Kitcher 2009). Yet, rather than calling for greater impartiality in biomedical research, feminists have often responded to con-

K. Intemann (✉)
Montana State University, Bozeman, MT, USA
e-mail: intemann@montana.edu

I. de Melo-Martín
Weill Cornell Medical College, Cornell University, New York, NY, USA
e-mail: imd2001@med.cornell.edu

M.C. Amoretti and N. Vassallo (eds.), *Meta-Philosophical Reflection on Feminist Philosophies of Science*, Boston Studies in the Philosophy and History of Science 317, DOI 10.1007/978-3-319-26348-9_5

cerns about commercial interests by calling for the incorporation of feminist values in both shaping research agendas and guiding scientific reasoning (Borgerson 2011; de Melo-Martin and Intemann 2011; Goldenberg 2013; Harding 2008; Intemann and de Melo-Martin 2014; Kourany 2010). Thus, the concern is that if commercial values can lead to bias because they fail to be impartial, then feminist values might also lead to bias in virtue of a similar failing.

We argue that this apparent tension is grounded on an equivocation about the sense in which values might be said to be problematically partial. With the grounds for the apparent tension dismantled, we show that feminists can resolve the apparent bias paradox and can also defend the claim that feminist values offer important resources for addressing the problems related to commercial interests.

5.2 The Bias Paradox in Biomedical Research

Feminist theorists are often interpreted as making the following three claims, which are thought to give rise to the bias paradox:

1. Commercial values ought not to influence biomedical research because such values are partial to some non-epistemic interests.
2. Feminist values are partial to some non-epistemic interests.
3. Feminist values ought to influence biomedical research.

Non-feminist approaches reject the third claim, seeing commercial and feminist values as equally problematic in scientific research. This is because it is traditionally thought that non-epistemic values of any kind have the potential to conflict with the epistemic obligations of scientists. That is, social, political, and economic values are normative ideals concerned with the way we want the world to be, rather than the way the world actually is. Thus, some believe that allowing such values to influence scientific reasoning is likely to lead to wishful thinking (Gross 1994; Haack 1998). In order to protect the epistemic integrity of science then, some scholars contend that scientific reasoning ought to be free of such values.

Along these lines, some have argued that promoting impartiality as a norm for individual scientists is necessary for protecting biomedical research from the influence of commercial values (Adam 2008; Krimsky 2003; Resnik 2007). That is, scientists should aim to refrain from endorsing any non-epistemic (social, political, or economic) values while engaging in scientific reasoning. While many acknowledge this may be difficult to do, they argue that impartiality is a norm that can and ought to be approximated. Striving for individual impartiality would presumably prevent commercial bias by making researchers more attentive to the ways in which financial interests might influence scientific judgments when conducting research. Researchers can thus try to ensure that financial interests do not inappropriately affect the framing of the questions, the selection and interpretation of evidence, the choice of methodologies, or the reporting of results. But the norm of impartiality might also help researchers be more attentive to the ways in which financial interests can lead to bias related to the direction of research. Impartial scientists might thus be less likely to conduct research that is merely profit-driven and that fails to be attentive to important public needs.

Yet feminist theorists have rejected this conception of impartiality for two main reasons. First, they have argued that this ideal of value-neutrality is unattainable even by approximation (Code 1991; Harding 1986; Longino 1990; Nelson 1990). The norm of individual impartiality seems to require researchers to evaluate whether financial interests are inappropriately affecting their investigations and to regulate themselves. But, a significant danger of financial interests is their ability to lead to *unconscious* bias. That is, even well-intentioned scientists may inadvertently allow their financial interests to influence how research questions are framed, which methodologies are preferred, and how data is selected and interpreted (Longino 1990, 2002; Nelson 1990; Solomon 2001). Such biases are nearly impossible for individual scientists to identify in their own research. This is so, precisely because they occur unintentionally, even when conscientious scientists believe they are rigorously adhering to the accepted methodologies of their field (Cain and Detsky 2008; Katz et al. 2003). For this reason, feminist theorists have often argued that knowledge is social, and that the appropriate locus of objectivity is at the level of communities, rather than individuals (Longino 1990; Nelson 1990).

Second, feminist philosophers have tended to argue that even if value-neutrality and disinterestedness were possible, they would not be desirable (Anderson 2004; Harding 2004; Kourany 2003). Some have argued that value judgments are *relevant* to scientific decision-making such that it would be irresponsible to ignore them (Anderson 2004; Douglas 2009; Dupré 2001; Intemann and de Melo-Martin 2010; Wylie and Nelson 2007). Moreover, many feminist scholars have pointed out that the aims of biomedical research are partly constituted by social and ethical aims (Dupré 2001; Harding 2008; Intemann and de Melo-Martin 2010; Kourany 2010). That is, biomedical research is concerned not only with producing true theories about the world, but also with saving lives, promoting human wellbeing, producing interventions that will be accessible and affordable for patients, and reducing or eliminating health disparities (Kourany 2010). This suggests that biomedical research is governed by multiple epistemological, pragmatic, and ethical aims and that researchers must consider which research questions, methodologies, and interventions would best promote those aims. As a result, this requires scientists to make value judgments in a way that would be incompatible with impartiality or disinterestedness.

How might feminist philosophers of science then be able to maintain the third proposition of the triad and also reject the use of commercial values in scientific reasoning? In the next section we show that the paradox is grounded on an equivocation about what makes the partiality of commercial values problematic.

5.3 On the Partiality of Commercial Values

It is worth pointing out that some might deny that the growing presence of commercial interests in science is problematic at all. Indeed, many would readily grant that a great deal of important medical research might not be done at all if it were not for the participation of private industry. Moreover, some commercially driven research has provided resources that are quite socially responsive. For instance, the industry's role was crucial for the development of statins, a class of drugs that has led to a significant reduction in cardiovascular disease morbidity and mortality (Zycher et al. 2010). In addition, it is not clear that profit-making aims are necessarily at odds with other traditional epistemic aims of research. Presumably, profit-making requires not just maintaining a good public image but also producing drugs and other interventions that actually work so that they will be prescribed and people will buy them. This requires epistemically sound evidence that such interventions are effective and safe.

Yet, despite the potential benefits of industry-funded research, there are persistent worries about the influence of commercial values in biomedical research. There are two main reasons why such influence is thought to be problematic. First, increasingly research is performed by scientists who receive funding, own stock, or collect honorariums or consulting fees from companies who are likely to profit from research outcomes. Thus, there is concern that such financial interests create conflicts of interest that can bias scientific reasoning. Several lines of evidence lend support to these concerns. For instance, as mentioned, research funded by industry is significantly more likely to report positive outcomes than studies funded by not-for-profit organizations (Bourgeois et al. 2010; Khan et al. 2008; Lundh et al. 2012; Nkansah et al. 2009; Sismondo 2008). Furthermore, evidence suggests that in the absence of statistically significant primary outcomes, industry funded researchers are more likely to look for and report positive subgroup findings than in non-industry funded trials (Sun et al. 2011). Studies have also found that industry supported reviews of drugs tend to be less attentive to methodological limitations of the included trials than the corresponding independent Cochrane reviews (Jørgensen et al. 2006; Tricco et al. 2009).

A second reason why financial interests may bias biomedical research is related to the influence that such interests have when setting research questions and priorities. Because industry significantly funds biomedical research, they have considerable power to determine the kinds of questions investigated (Angell 2004), as well as how problems and solutions are framed (Intemann and de Melo-Martin 2010). This can shape research programs and priorities in a variety of ways. For example, because biomedical companies need products that can offset the expenses of drug and device research and development, they prefer strategies that minimize economic risks. This might include combining agents that are already approved or making slight chemical changes in order to extend patent rights on particular drugs, even though these changes in most cases bring little or no added benefit to patients (Brody 2007). Additionally, financial interests incentivize research that targets populations with the most resources to pay for drugs. Many of the diseases that affect the world's poorest people, and that constitute a significant burden of disease (WHO 2008), are neglected. Well documented is the "10/90 gap," whereby only 10 % of worldwide expenditure on health research and development is devoted to the problems that affect primarily the poorest 90 % of the world's population (Currat et al. 2004; Kilama 2009; Stevens 2008; Vidyasagar 2006). Even in cases where research is geared toward addressing a pressing public health need, there are concerns that financial interests lead to pursuing interventions that are patentable and profitable, even when other kinds of solutions could be more effective (Colagiuri et al. 2006; Magnusson 2009; Yach et al. 2004). For example, lifestyle changes or preventative interventions might have a more significant effect on reducing the burden of diabetes, obesity, and hypertension than the use of particular drugs, but these alternatives might be less likely to produce profits.

Hence, concerns that commercial values and interests can negatively influence biomedical research are not unjustified. Moreover, such worries are likely to be particularly concerning to feminists given their political commitments to reduce health inequalities and improve the conditions of marginalized groups. But, insofar as feminists maintain that *some values*, namely feminist values, ought to influence scientific reasoning, what grounds do feminists have for criticizing the use of commercial values in biomedical research?

Concerns that feminist approaches succumb to the bias paradox are grounded on the assumption that such approaches take commercial interests to be problematic because they are *partial*. What exactly does this mean? In a trivial sense, values might be said to be partial in that they endorse or promote certain ends or interests over others, e.g., promoting profit interests over non-profit related interests. If this were what feminists claim, then it is easy to see how the bias paradox would arise, because feminist values also obviously promote certain ends over others. But this cannot be why commercial interests are thought to be problematic. All values are partial in this trivial sense. Even epistemic values, such as truth, empirical adequacy, consistency, or explanatory power, would be partial in this sense and no one seems to claim that appealing to these values would be problematic.

One might argue that commercial values are problematic because they are partial in the sense that they prioritize non-epistemic interests over epistemic ones. Of course, to the extent that commercial values play a role that conflicts with the epistemic goals of the research, then such values are problematic. Yet, as mentioned above, it is not clear that commercial interests necessarily conflict with epistemic aims, as advancing commercial interests often requires epistemically sound research. Even more importantly, this does not account for concerns related to biased research agenda, as such research may be epistemically flawless even when it fails to address pressing public health needs or improve human well-being in substantive ways.

Instead, feminists' objections to commercial values are best understood as a worry about the *dominant* role that such values have in the field of biomedical research, often to the exclusion of other values and interests that may be at stake. That is, worries about commercial interests stem from the disproportionate influence that such values have in setting research agendas, developing interventions, determining methodologies, and disseminating (or refrain from disseminating) information. Profit-driven interests are not representative of the interests of all stakeholders affected by research, yet they are given more weight because of the resources that their advocates possess. As a result, such values often trump other relevant epistemic and social values for reasons that would not be widely accepted. This is the relevant sense in which feminists take commercial values to be unacceptably partial. In the next section, we will consider whether feminist values can be said to be partial in the same way (Solomon 2012).

5.4 On the Partiality of Feminist Values

While many feminist theorists have argued that scientific reasoning and practices ought to incorporate more feminist values and interests, there is disagreement about what this means or requires (Anderson 2004; Harding 2008; Longino 1990; Nelson 1990; Wylie and Nelson 2007). For our purposes, we take feminist values to be those aimed at reducing power inequalities and at challenging systems of oppression. What particular values this might consist in is a matter of debate. In biomedical research, for example, feminists might be broadly committed to reducing health disparities and addressing the health needs of marginalized groups. There might be a variety of ways of understanding who counts as marginalized: those who are economically and socially least well-off, those who are most at-risk for a certain disease; those who suffer from diseases and conditions that seriously impair human flourishing. Similarly, there may be disagreements about how equality is best understood or promoted. While these are important questions, our concern is only to consider whether feminist values, broadly construed, are partial in the same sense as commercial values seem to be. Here we identify two different roles that feminist values might play in biomedical research and show that neither of them commits feminists to the position that feminist values are partial in the same problematic sense that commercial values have been partial in biomedical research.

5.4.1 Feminist Values as Promoting Procedural Impartiality

Some have argued that feminist values have an important role to play in structuring scientific communities and providing mechanisms that balance or minimize the negative effects of the personal interests of individual scientists (Longino 1990, 2002; Solomon 2001). Longino, for example, argues that objective scientific communities are comprised of community of inquirers with diverse values and interests who have equal intellectual authority to challenge and scrutinize scientific research. If adequate avenues for public criticism exist, such as transparency about methodologies and peer review, then idiosyncratic values and interests influencing scientific decision-making are likely to be caught and corrected by others in the community who have different values and interests. These mechanisms do not make values disappear but they ensure that values receive critical scrutiny and that any inadequate influence such values have on scientific reasoning will be identified and corrected. Thus, on this account, commercial values, or indeed other values that conflict with feminists ones such as sexist or racist values, need not be eliminated from biomedical research. What is required is that scientific communities are formed by people with a variety of values and interests and that mechanisms exist to ensure that all such values are open to consideration and criticism on equal terms. On this account then, the inclusion of feminist and other non-commercial values will be particularly important because these values have been historically under-represented or excluded from scientific inquiry. Yet, it is not that these values are to be given less scrutiny than commercial values. The requirement is thus one of *procedural* impartiality. While individual scientists need not be free of the influence of values, the procedures by which values and other background assumptions are evaluated are fair and impartial to the content of those values. Much like the rules of a game are designed to treat players impartially so as to not given any particular team an advantage, procedural impartiality gives all values an equal chance to compete.

But if feminist values do not trump other relevant epistemic and social values, then this view does not entail that feminist values are partial in the same inappropriate way that feminist take commercial values to be. Claims about the problematic influence of feminist values on grounds that they are simply *partial* are thus undermined.

In addition to dissolving the bias paradox, this account of procedural impartiality provides several potential resources for helping to address the ways in which financial interests may lead to bias in biomedical research. As mentioned earlier, procedural impartiality involves the use of a variety of mechanisms aimed to facilitate the identification and correction of problematic background assumptions that might reflect financial interests and can lead to unsafe or ineffective medical interventions. In addition, procedural impartiality requires scientific communities to be comprised of members with diverse interests and values, which includes inquirers with different conceptions of the good life and the public good, as well as with diverse aims in conducting biomedical research. As a result, scientific communities are likely to pursue a wider range of hypotheses and of medical interventions, other than those geared solely or mainly toward commercial interests. This would make it more likely that the research agenda would be at least partly shaped by those attempting to address pressing public needs.

Similarly, because scientific communities would incorporate researchers with various values and interests, they would frame research problems in different ways and would utilize a variety of methodological approaches. For example, some HIV vaccine researchers might aim to produce a vaccine that is causally efficacious in preventing infection from a broad range of HIV strains. They might thus design highly controlled trials able to identify causal mechanisms. Other vaccine researchers might direct their efforts to create an intervention that is able to decrease

HIV infection in resource-poor areas where there is less access to medical facilities and more risk of HIV infection and where evidence of efficaciousness and effectiveness might not coincide (Cartwright 2006; Intemann and de Melo-Martin 2010). This might justify utilizing an adaptive clinical trial design that allows methodological changes during the trial in response to new data or paying attention to the social and cultural aspects that might make adherence of some HIV preventive strategies difficult (Intemann and de Melo-Martin 2010; Woodsong et al. 2013). This sort of epistemological and methodological pluralism helps to highlight strengths and weakness of the various approaches and different epistemic tradeoffs. Thus, it could be useful for both identifying the limitations of certain profit-driven approaches, while also generating alternative conceptualizations of a problem and different methodological approaches.

This approach, then, not only avoids the apparent bias paradox, but also has important resources for attempting to address some of the problems related to the commercialization of science. It is an approach that has been developed primarily by feminist empiricists who have sought to correct for the historical absence of feminist values and interests in research (Borgerson 2011; Longino 1990; Nelson 1990). In doing so, it provides resources for offsetting the undue influence that commercial interests (as well as other values held by those with more power) have had in biomedical research.

5.4.2 Feminist Values as Contributing to Strong Objectivity

Some feminists, however, have made the stronger claim that feminist values should not only be included among those values that operate in biomedical research, but that they should be given greater weight or priority than other values such as commercial values (Goldenberg 2013; Harding 2008; Intemann and de Melo-Martin 2014; Kourany 2010). Standpoint feminists, for example, have argued that we need to reconceptualize scientific objectivity as *strong objectivity*, which requires not that science be free of values, but rather that scientific practices be structured so as to limit ways in which those in power can influence, shape, and distort the production of knowledge (de Melo-Martin and Intemann 2011; Harding 2008). This approach argues that certain social and ethical goals should be recognized as partly constitutive of the aims of biomedical research. That is, biomedical research should be concerned with producing knowledge that is relevant to marginalized groups in order to counteract, eliminate, or minimize the ways in which oppressive systems are limiting their health, wellbeing, or life prospects. Kourany, for instance, has argued that all scientists ought to endorse egalitarian values in setting research agendas and conducting research (Kourany 2003, 2010). Others assert that researchers should adopt a normative commitment to examining scientific phenomena in ways that challenge, rather than reinforce, oppressive systems (de Melo-Martin and Intemann 2011; Harding 1986, 2008; Hundleby 1997; Intemann 2010). This requires investigating public health problems by starting from the conditions and interests of marginalized groups so as to produce knowledge and medical interventions that will be more likely to benefit them (Crasnow 2006; de Melo-Martin and Intemann 2011).

On this account, feminist values would play a more prominent role and be given more weight than commercial values. For instance, the interests of marginalized groups would have priority in terms of setting the agenda for biomedical research. Moreover, these interests and values would play a role in scientific reasoning. Consider, for example, a research group aiming to prevent cervical cancer. Such a group would share a commitment to significantly lowering cervical cancer morbidity in ways that do not reinforce or exacerbate existing social and political inequalities. This means examining how unjust social conditions might contribute to cervical cancer as a health problem and how such conditions might also pose challenges to potential medical interventions. This research would begin by inquiring who is the most at risk for cervical cancer, and what type of medical interventions might work best for that population, given the existing social, economic, and political constraints. For example, researchers may ask whether there are reasons to think that focusing on behavioral modifications, such as trying to prevent HPV infections by using condoms, might be more cost-effective and accessible to those in resource-poor areas, or whether there are there social and cultural challenges to such strategies, such as women having less power to negotiate the terms of sexual encounters. Such a research group would also be obligated to consider how HPV and/or cervical cancer affect or manifest differently in marginalized groups, e.g., non-whites, gay/lesbian/bisexual/transgendered, those from resource-poor contexts. When designing a clinical trial, researchers would consider the social aims of the research to help inform methodological decisions such as appropriate clinical trial site, what constitutes a representative subject pool, or what the best duration of the trial might be.

On this account, feminist values may be partial not only in the unproblematic sense of endorsing some values or aims, but in the stronger sense of promoting feminist values over commercial and other interests that conflict with feminist values. Thus, one might worry that this stronger claim is vulnerable to the bias paradox.

There are two reasons, however, to think that even on this view, feminist values are not partial in the same, unacceptable way as commercial values are. First, although, as we have said, commercial values need not be in conflict with epistemic ones, the history of research funded by tobacco companies or the more recent case of Vioxx show that at least some times commercial interests do in fact conflict with the production of objective research (McGarity and Wagner 2008; Michaels 2008). Feminist values, however, are unlikely to trump or come into conflict with epistemic aims in the ways that commercial values can. It is not simply that successfully addressing public health needs and challenging oppressive systems will require medical and other interventions that are in fact effective. Attempting to produce research guided by feminist values that simply appears to be effective by in fact is not would be of little use to feminists. Research that merely appears to be objective is unlikely to combat oppression, help the marginalized or address public health concerns. However, as research funded by tobacco companies, for instance, has shown, that is certainly not the case for research that is influenced by commercial values. Vioxx earned Merk billions of dollars, and harmed hundreds of patients, before the drug was taken off the market (James et al. 2007). Feminist values, on the other hand, are inconsistent with the production of science that is unreliable. As the cases of biased research resulting from the influence of financial interest indicates, commercial values can be consistent with ignoring or actively opposing the epistemic goals of research. It is not clear at all, however, what it would mean to conduct research under feminist commitments that results in biased research in virtue of such commitments.

Second, and equally important, this approach does not require that researchers neglect the interests and values of any stakeholder affected by the research. While the needs and interests of marginalized groups may be said to have priority, this weight is not accorded disproportionately on some arbitrary basis. Instead, this commitment is morally justified by taking into account all of those affected by research. Scientific knowledge is a resource for understanding and addressing social problems, generating interventions, and developing public policies, with great potential to address social injustices and improve lives (Kourany 2003, 2010). As such, it is a public good that arguably ought to be used equitably and in particular one that should benefit those who are least well off (Daniels 2008; Kitcher 2011; Kourany 2010; Pogge 2002). The interests of marginalized groups have priority in this sense because they are disproportionately affected by health, environmental, and social problems in virtue of their material circumstances, the nature of their countries' economies, the lack of resources they possess to mitigate these problems, and the existence of inequalities. Moreover, it is often the case, in climate change for instance, that marginalized groups bear less responsibility for the creation of these problems. Thus, there are compelling ethical reasons to adopt the political commitments advocated on this feminist approach and to ensure that scientific research is consistent with such commitments.

Those who claim that feminist values ought to be prioritized in biomedical research do so, then, on the grounds that partiality towards feminist values is more justified than partiality towards commercial values. It is not simply that the partiality of commercial values is problematic because they are likely to trump other relevant values and interest. It is also that there are no good reasons for giving commercial values priority in biomedical research. Doing so would be particularly problematic given the other interests at stake and the moral significance those interests arguably have.

Of course, this does not mean that commercial interests would need to be eliminated from biomedical research altogether, which, practically speaking, would seem quite difficult. There may be cases when commercial values and feminist values coincide. As mentioned, industry provides a significant amount of biomedical research funding and eliminating profit incentives may eliminate these resources. But, in this case, the justification for the incorporation of profit-driven values must be that the use of such values will produce research that is socially responsive and particularly that attends to the needs of the least well-off. That is, profit-driven research would be justified insofar as it also promotes the well-justified feminist aims.

As with the previous interpretation, this account of the role of feminist values in research is able to dissolve the bias paradox. Similarly, this approach also provides important resources for addressing the problems posed by commercial interests.

Like procedural impartiality, standpoint theorists also advocate for diversity in scientific communities, but in this case the diversity that is important is the participation of those from different social groups, both within research groups and among those advising research groups. A diverse scientific community will bring different life experiences, interests, and values, all of them sharing a defeasible normative commitment to challenging systems of oppression, which will be useful for generating a broad range of hypotheses, strategies, methods, interventions and technologies thought to be appropriate solutions to a particular problem. Such diversity will also be useful in identifying and correcting problematic background assumptions. This feminist account then is well-equipped to deal with threats to the epistemic soundness of research.

However, because of this account's critical approach to power structures, a scientifically diverse community needs to recognize an important ethical and political imperative to conduct research in ways that challenge, rather than reinforce systems of oppression, at least in areas of research that have social implications. To the extent that some research is merely profit-driven, or is intended to primarily serve the interests of those standing to profit, such research will be at least prima facie problematic. Similarly, values or interests that are likely to further oppressive conditions can be justifiably excluded from playing a role with relation to hypothesis proposal, solutions considered, or questions asked.

The ethical and political commitment to produce epistemically sound research that is socially responsive allows this feminist view to scrutinize background assumptions, research questions, and methodological choices to determine not just whether they are epistemically appropriate but whether they promote pressing public health needs. On this view, the commitments to challenge systems of oppression gives feminists resources when considering what problems to attend to, how to frame those problems, and what kind of solutions are thought to be appropriate. Thus, scientific communities structured according to these ideals will provide a pluralist research agenda and one that will be particularly attentive to the 10/90 gap for instance.

5.5 Dissolving the Bias Paradox: Some Conclusions

We have argued that the bias paradox trades on an ambiguity about what it means for values to be partial in an unacceptable way. A value might be partial in the unproblematic sense that it seeks to promote some aims over others. In this sense, feminist values, like commercial values, are partial, but so are epistemic values. The partiality of commercial values however is problematic when commercial values arbitrarily promote some interests to the exclusion of other relevant values and interests. In the case of biomedical research, the concern is that commercial values dominate research solely because of the power their proponents have and not because they are supported by good reasons that would be widely accepted. As a result, they have the potential to bias both the direction and content of biomedical research in ways that can be epistemically unsound and that are arguably ethically suspect.

If this is correct then, the apparently inconsistent triad can be revised so as to make the ambiguity present explicit:

1. Commercial values are problematic in biomedical research insofar as they arbitrarily promote economic interests to the exclusion of epistemic or social values justifiably held by stakeholders affected by the research.
2. Feminist values in biomedical research aim to reduce power inequalities and promote the epistemic and social interests of all stakeholders in ways that would be widely accepted as justified.
3. Feminist values ought to influence biomedical research.

While feminists may disagree about how feminist political and ethical commitments ought to be understood in particular research contexts, as well as the roles it is appropriate for such values to play, we have argued that the partiality of feminist values is not problematic in the same way as that of commercial values is. Feminists who claim that feminist values have a role to play in diversifying the values and interests advanced in science are not guilty of being partial to feminist values over others that might be represented in scientific communities. Even feminists who make the stronger claim that feminist values ought to be given more weight in biomedical research argue that there are non-arbitrary grounds for doing so. On this view, what is problematic about commercial values is not that they involve partiality or values per se, but rather that they involve value judgments that are unjustified.

References

Adam, M. (2008). Promoting disinterestedness or making use of bias? In M. Carrier, D. Howard, & J. Kourany (Eds.), *The challenge of the social and the pressure of practice* (pp. 235–254). Pittsburgh, PA: University of Pittsburgh Press.

Anderson, E. (2004). Uses of value judgments in science: A general argument, with lessons from a case study of feminist research on divorce. *Hypatia, 19*(1), 1–24.

Angell, M. (2004). *The truth about the drug companies: How they deceive us and what to do about it*. New York: Random House.

Antony, L. (1993). Quine as feminist: The radical import of naturalized epistemology. In L. Antony & C. Witt (Eds.), *A mind of one's own* (pp. 110–153). Boulder, CO: Westview.

Borgerson, K. (2011). Amending and defending critical contextual empiricism. *European Journal for Philosophy of Science, 1*(3), 435–449.

Bourgeois, F. T., Murthy, S., & Mandl, K. D. (2010). Outcome reporting among drug trials registered in ClinicalTrials.gov. *Annals of Internal Medicine, 153*(3), 158–166.

Brody, H. (2007). *Hooked: Ethics, the medical profession, and the pharmaceutical industry*. Lanham: Rowman & Littlefield.
Cain, D. M., & Detsky, A. S. (2008). Everyone's a little bit biased (even physicians). *Jama-Journal of the American Medical Association, 299*(24), 2893–2895.
Cartwright, N. (2006). Well-ordered science: Evidence for use. *Philosophy of Science, 73*(5), 981–990.
Code, L. (1991). *What can she know? Feminist theory and the construction of knowledge*. Ithaca: Cornell University Press.
Colagiuri, R., Colagiuri, S., Yach, D., & Pramming, S. (2006). The answer to diabetes prevention: Science, surgery, service delivery, or social policy? *American Journal of Public Health, 96*(9), 1562–1569.
Crasnow, S. (2006). Feminist anthropology and sociology: Issues for social science. In S. Turner & M. Risjord (Eds.), *Handbook of the philosophy of science* (Vol. 15, pp. 827–860)., Philosophy of anthropology and sociology Amsterdam: Elsevier.
Currat, L., de Francisco, A., Al-Tuwaijri, S., Ghaffar, A., & Jupp, S. (2004). *10/90 report on health research 2003–2004*. Geneva: Global Forum for Health Research.
Daniels, N. (2008). *Just health: Meeting health needs fairly*. Cambridge: Cambridge University Press.
de Melo-Martin, I., & Intemann, K. (2011). Feminist resources for biomedical research: Lessons from the HPV vaccines. *Hypatia, 26*(1), 79–101.
De Winter, J. (2012). How to make the research agenda in the health sciences less distorted. *Theoria, 27*, 75–93.
Douglas, H. E. (2009). *Science, policy, and the value-free ideal*. Pittsburgh: University of Pittsburgh Press.
Dupré, J. (2001). *Human nature and the limits of science*. New York: Oxford University Press.
Fausto-Sterling, A. (1992). *Myths of gender: Biological theories about women and men* (2nd ed.). New York, NY: BasicBooks.
Goldenberg, M. (2013). How can feminist theories of evidence assist clinical reasoning and decision-making? *Social Epistemology: A Journal of Knowledge, Culture and Policy*. doi:10.1080/02691728.2013.794871
Gross, B. (1994). What could a feminist science be? *Monist, 77*(4), 434–444.
Haack, S. (1998). *Manifesto of a passionate moderate: Unfashionable essays*. Chicago: University of Chicago Press.
Harding, S. G. (2004). A socially relevant philosophy of science? *Resources from Standpoint Theory's Controversiality. Hypatia, 19*(1), 25–47.
Harding, S. G. (1986). *The science question in feminism*. Ithaca: Cornell University Press.
Harding, S. G. (2008). *Sciences from below: Feminisms, postcolonialities, and modernities*. Durham: Duke University Press.
Hundleby, C. (1997). Where standpoint stands now. *Women & Politics, 18*(3), 25–43.
Intemann, K. (2010). Sciences from below: Feminisms, postcolonialities, and modernities. *Hypatia, 25*(2), 464–469.
Intemann, K., & de Melo-Martin, I. (2010). Social values and scientific evidence: The case of the hpv vaccines. *Biology and Philosophy, 25*(2), 203–213.
Intemann, K., & de Melo-Martin, I. (2014). Addressing problems in profit-driven research: How can feminist conceptions of objectivity help? *European Journal For Philosophy of Science, 4*(2), 135–151.
James, M. J., Cook-Johnson, R. J., & Cleland, L. G. (2007). Selective COX-2 inhibitors, eicosanoid synthesis and clinical outcomes: A case study of system failure. *Lipids, 42*(9), 779–785.
Jørgensen, A. W., Hilden, J., & Gøtzsche, P. C. (2006). Cochrane reviews compared with industry supported meta-analyses and other meta-analyses of the same drugs: Systematic review. *BMJ, 333*(7572), 782.
Katz, D., Caplan, A. L., & Merz, J. F. (2003). All gifts large and small—toward an understanding of the ethics of pharmaceutical industry gift-giving. *American Journal of Bioethics, 3*(3), 39–46.
Keller, E. F. (1985). *Reflections on gender and science*. New Haven: Yale University Press.

Khan, S. N., Mermer, M. J., Myers, E., & Sandhu, H. S. (2008). The roles of funding source, clinical trial outcome, and quality of reporting in orthopedic surgery literature. *American Journal of Orthopedics, 37*(12), E205–E212. discussion E212.
Kilama, W. L. (2009). The 10/90 gap in sub-saharan Africa: Resolving inequities in health research. *Acta Tropica, 112*(Suppl. 1), S8–S15.
Kitcher, P. (2011). *Science in a democratic society*. Amherst, NY: Prometheus Books.
Kourany, J. A. (2003). A philosophy of science for the twenty-first century. *Philosophy of Science, 70*(1), 1–14.
Kourany, J. A. (2010). *Philosophy of science after feminism*. Oxford: Oxford University Press.
Krimsky, S. (2003). *Science in the private interest: Has the lure of profits corrupted biomedical research?*. Lanham: Rowman & Littlefield.
Longino, H. E. (1990). *Science as social knowledge: Values and objectivity in scientific inquiry*. Princeton, NJ: Princeton University Press.
Longino, H. E. (2002). *The fate of knowledge*. Princeton, NJ: Princeton University Press.
Lundh, A., Sismondo, S., Lexchin, J., Busuioc, O. A., & Bero, L. (2012). Industry sponsorship and research outcome. *Cochrane Database of Systematic Reviews, 12*, MR000033.
Magnusson, R. S. (2009). Rethinking global health challenges: Towards a "global compact" for reducing the burden of chronic disease. *Public Health, 123*(3), 265–274.
McGarity, T. O., & Wagner, W. (2008). *Bending science: How special interests corrupt public health research*. Cambridge, MA: Harvard University Press.
Michaels, D. (2008). *Doubt is their product: How industry's assault on science threatens your health*. Oxford: Oxford University Press.
Nelson, L. H. (1990). *Who knows: From Quine to a feminist empiricism*. Philadelphia: Temple University Press.
Nkansah, N., Nguyen, T., Iraninezhad, H., & Bero, L. (2009). Randomized trials assessing calcium supplementation in healthy children: Relationship between industry sponsorship and study outcomes. *Public Health Nutrition, 12*(10), 1931–1937.
Pogge, T. W. M. (2009). The health impact fund and its justification by appeal to human rights. *Journal of Social Philosophy, 40*(4), 542–569.
Pogge, T. W. M. (2002). *World poverty and human rights: Cosmopolitan responsibilities and reforms*. Cambridge, MA: Polity.
Reiss, J., & Kitcher, P. (2009). Biomedical research, neglected diseases, and well-ordered science. *Theoria, 24*, 263–282.
Resnik, D. B. (2007). *The price of truth: How money affects the norms of science*. Oxford: Oxford University Press.
Sismondo, S. (2008). Pharmaceutical company funding and its consequences: A qualitative systematic review. *Contemporary Clinical Trials, 29*(2), 109–113.
Solomon, M. (2001). *Social empiricism*. Cambridge, MA: MIT Press.
Solomon, M. (2012). The web of valief. An assessment of feminist radical empiricism. In S. L. Crasnow & A. M. Superson (Eds.), *Out from the shadows: Analytical feminist contributions to traditional philosophy* (pp. 435–450). New York, NY: Oxford University Press.
Stevens, P. (2008). Diseases of poverty and the 10/90 gap. In P. Stevens (Ed.), *Fighting the diseases of poverty* (pp. 126–140). New Brunswick, NJ: Transaction.
Sun, X., Briel, M., Busse, J. W., You, J. J., Akl, E. A., Mejza, F., et al. (2011). The influence of study characteristics on reporting of subgroup analyses in randomised controlled trials: Systematic review. *BMJ, 342*, d1569.
Tricco, A. C., Tetzlaff, J., Pham, B., Brehaut, J., & Moher, D. (2009). Non-cochrane vs. Cochrane reviews were twice as likely to have positive conclusion statements: Cross-sectional study. *Journal of Clinical Epidemiology, 62*(4), 380–386.
Vidyasagar, D. (2006). Global notes: The 10/90 gap disparities in global health research. *Journal of Perinatology, 26*(1), 55–56.
WHO. (2008). *The global burden of disease: 2004 update*. Geneva: World Health Organization.

Woodsong, C., MacQueen, K., Amico, K. R., Friedland, B., Gafos, M., Mansoor, L., et al. (2013). Microbicide clinical trial adherence: Insights for introduction. *Journal of the International AIDS Society, 16*, 18505.

Wylie, A., & Nelson, L. H. (2007). Coming to terms with the values of science: Insights from feminist science studies scholarship. In H. Kincaid, J. Dupre, & A. Wylie (Eds.), *Value-free science? Ideals and illusions* (pp. 58–86). Oxford: Oxford University Press.

Yach, D., Hawkes, C., Gould, C. L., & Hofman, K. J. (2004). The global burden of chronic diseases: Overcoming impediments to prevention and control. *JAMA, 291*(21), 2616–2622.

Zycher, B., DiMasi, J., & Milne, C. (2010). Private sector contributions to pharmaceutical science: Thirty-five summary case histories. *American Journal of Therapeutics, 17*(1), 101–120.

Chapter 6
Values and Evidence in Feminist Philosophy and in Neuroscience

Robyn Bluhm

6.1 Introduction

Research on neuroscience was one of the earliest, and remains one of the most extensive, areas of feminist science criticism. In the 1970s and 1980s, a number of feminist scholars in neuroscience, psychology, philosophy, and other disciplines began to criticize research that purported to show cognitive differences between women and men and to trace these differences back to the structure of the brain. These critics found, through a careful analysis of studies that claimed to show sex differences in the brain, that the conclusions drawn in this research were frequently not supported by the evidence. Moreover, by linking the methods and the kinds of data collected in these sciences to prevailing social stereotypes, the feminist critics of this research also showed that the scientists' conclusions owed more to prevailing beliefs about gender than to the evidence itself.

Recently, a new spate of books and articles have appeared that take much the same approach to current research on sex differences as did the work published several decades earlier (Kaiser et al. 2009; Fine 2010; Jordan-Young 2010; Meynell 2012; Bluhm 2013a, b). This recent body of research shows that scientists investigating sex differences still appear to be using much the same methods, drawing much the same inferences, and relying just as much on gender stereotypes instead of solid evidence to ground their conclusions about sex differences in the brain and in behavior. To borrow Fine's (2010) term, "neurosexism" is flourishing in contemporary science.

In addition to its striking similarity to earlier criticisms, another interesting feature of this second era of feminist neuroscience criticism is how little it draws on feminist philosophy of science, even when it is conducted by philosophers well aware of this literature. This is surprising, since much of the foundational early work in feminist

R. Bluhm (✉)
Michigan State University, East Lansing, MI, USA
e-mail: rbluhm@msu.edu

M.C. Amoretti and N. Vassallo (eds.), *Meta-Philosophical Reflection on Feminist Philosophies of Science*, Boston Studies in the Philosophy and History of Science 317, DOI 10.1007/978-3-319-26348-9_6

philosophy of science drew extensively on feminist criticisms of neuroscience. For example, Longino (1990), Nelson (1990, 1995), and Harding (1986, 1993) all develop their philosophies of science in part by analyzing the work of such feminist critics of science as Bleier (1984, 1986), Fausto-Sterling (1985), and Birke (1986).

A closer look at this early feminist philosophy of science shows, however, that it was much less critical of sex difference research than were the feminist scientists whose work they discussed. One of the central aims of this paper is to understand why this is so. In investigating this question, I will begin by briefly describing the conceptual framework that underlies much research on sex/gender differences in the brain. "Brain organization theory" posits that cognitive, behavioral, and personality differences between women and men can be traced back to the influence of sex hormones on the brain during fetal development. I will then briefly discuss feminist criticisms of this research, using Ruth Bleier's work as a representative example, and will then show that feminist philosophers of science backed away from the full force of Bleier's criticisms.

I will argue that feminist philosophy of science *can* support a deeper critique. The second aim of this paper is to identify resources and approaches that can help us to do so. In making this case, I will draw on work by Sharyn Clough and Maya Goldenberg, who argue that feminist philosophers of science should focus on developing accounts of how social issues and values interact with empirical evidence. Using examples from Rebecca Jordan-Young's recent empirical critique of brain organization, I conclude by sketching ways in which this approach could work.

6.2 Brain Organization Theory

The central framework underlying most neuroscience research on sex/gender differences is brain organization theory. The main idea is that exposure to different levels of sex hormones during fetal development permanently give the brain a male (or a female) structure. Moreover, because structural differences lead to differences in function, brain organization theory views fetal hormone exposure as the root cause of sex/gender differences in cognition, personality characteristics, interests, and sexual behavior. Early studies using brain organization theory tended to focus mostly on the effects of testosterone, viewing the female brain as the "default" that develops in the absence of testosterone exposure, but other sex hormones and their metabolic products have also come to be recognized as increasingly important.

Briefly, brain organization theory was originally inspired by work in rodents that showed that androgens produced by the testes suppress or block the ability of hypothalamic neurons to respond cyclically and therefore to regulate pituitary and gonadal cyclic function. Phoenix et al. (1959) conducted a study that has come to be regarded as the genesis of brain organization theory; they exposed female guinea pigs to increased testosterone levels during fetal development and found that, as adults, they displayed more "male" sexual behavior (i.e., mounting) and less "female" sexual behavior (i.e., lordosis) than unexposed female guinea pigs. Later studies examined the effects of testosterone on adult hormone cycles. Male rats that are deprived of androgens at birth (by castration) and then given estrogens as adults showed the patterns of cyclic hormone release that occur in female rats. In addition, female rats that were given a single androgen shot did not have female-characteristic hormone cycles when they reached maturity; for example, they did not ovulate (Harris and Levine 1965). The fact that a single intervention had permanent effects on later brain function suggested that early exposure to sex hormones has an "organizing" effect on the brain.

Findings in these rodent species further suggested that there might be similar "organizing" effects of sex hormones in human beings. Many researchers have conducted studies that examined individuals who were exposed to atypical levels of hormones during fetal development. In some cases, this involved higher or lower levels of hormones associated with their sex (i.e., testosterone for males and estrogen and progesterone for females); in others, this involved exposure to the "wrong" hormones for their sex. Researchers looked for evidence of subsequent effects on sexual orientation and other sexual behaviors, on cognitive abilities, and on personality characteristics.

One important manifestation of brain organization theory is its interaction with the idea that the left and right hemispheres of the brain have different structures and therefore different functions. It has been proposed that the reason that the brain is lateralized is because of exposure to hormones during fetal development. The lateralization hypothesis is intended to explain the relationships among a number of factors, including left-handedness (or right nondominance), immune disorders, learning disorders, and above average visuospatial function. All of these, however, have been found to be more common in males. The hypothesis for explaining the relationship among all of these characteristics, then, was differences in testosterone exposure during fetal development: "There is probably some influence that slows the growth of parts of the left hemisphere so that… the corresponding regions on the right develop relatively more rapidly" (Geschwind and Galaburda 1987: 11). Because female fetuses are also exposed to some testosterone, the left hemisphere also develops more slowly than the right, but higher levels of testosterone in males explain the more common occurrence in this sex of the factors listed above. Moreover, higher- or lower-than-average exposure to testosterone in males, or higher-than average exposure in females, explains the variability in these factors within each sex.

Much of Geschwind and Galaburda's discussion of lateralization is related to clinical effects, however, it has also been frequently used to explain cognitive differences between women and men. For example, Benbow and Stanley have used it to explain both the greater number of boys who have outstanding mathematical ability, but also average differences in mathematical achievement between boys and girls (e.g., Benbow and Stanley 1983). Neuroscience researchers also draw on the idea that there are sex differences in brain lateralization to explain purported differences in language processing between women and men (see the literature review in Kaiser et al. 2009).

I will discuss specific feminist criticisms of brain organization theory in the next section, but for now I want to point out one of its broader implications. Doell and Longino (1988) describe brain organization theory as promoting a "linear model" of development, on which genetic (chromosomal) differences lead to hormonal differences, which lead to brain differences, which in turn lead to behavioral differences. There is no room for nonlinear feedback in this process; at best, for example, the influence of behavior on hormone levels or on neural activity is viewed as minimal. As a result, the model tends to be allied with a strong biological determinism, on which sex/gender differences in behavior and cognition tend to be viewed as largely "natural" and unchangeable. While more recent versions of the theory allow for environmental effects, including differences in the social environments experienced by male and female children, these environmental differences are viewed as acting on brains that are already fundamentally shaped by hormone exposure. The environmental effects thus merely exacerbate the pre-existing hormone differences.

6.3 Feminist Criticisms of Brain Organization Theory

Although brain organization has been and continues to be very influential, it has also been a target of criticism. Some of the earliest critics of brain organization theory were feminist scientists trained in biology or neuroscience, who carefully analyzed the evidence for the theory by looking at the details of the studies that were supposed to support it. Their major conclusions were, first, that many of the that the specific hypotheses based on the theory, or even the broader theory itself, were not well-supported by the available evidence and, second, that in drawing their conclusions to the contrary, the researchers who were conducting these studies were influenced by the prevailing, largely sexist, social views on sex differences. It should be noted that these are two distinct criticisms, as one need not accept the feminist point in order to accept the empirical criticisms of the research. I will return to this point later in the chapter.

Similarly, and at around the same time, a number of feminist philosophers of science were beginning to look at sexism in science and at the way that broader social beliefs about women and men were deployed in research. Both of these groups aimed to judge the scientific research on empirical grounds. Yet, as I shall show, they came to rather different conclusions. This may have been, in part, because of their different disciplinary backgrounds, (neuro)science versus philosophy, which led them to emphasize different aspects of the research and the context within which it was conducted, but I will argue that it also was due to the *specific* philosophical questions to which the latter group attended. In this section, I will describe Ruth Bleier's critical assessment of research based on brain organization theory. I will then discuss the way that brain organization theory and these criticisms from feminist scientists were taken up by the philosophers Lynn Hankinson

Nelson and Helen Longino and argue that these latter authors did not give enough credit to the criticisms of the former.

Ruth Bleier opens her 1984 book *Science and Gender: A Critique of Biology and its Theories on Women* on a clearly feminist note; she describes science as playing a role "in the creation of an elaborate mythology of women's biological inferiority as an explanation for their subordinate position in the cultures of Western civilization" (Bleier 1984: vii). Yet much of her work in this book is based on empirical, rather than political, analysis; she aims to show that many studies that purport to demonstrate or to explain differences between women and men "have been methodologically and conceptually unsound and inconclusive" (p. viii). Neuroscience research occupies a central position in her critiques of biology. She devotes a chapter to describing brain development, in order to show that this development depends on input from the external world, making it impossible to separate biology/nature from culture/nurture. In the following chapter, she addresses research that links "sex" hormones and their influence the brain with a variety of gendered characteristics or areas of gender difference.

These criticisms bring her to the heart of brain organization theory. Recall that this theory began with a study that showed that fetal/neonatal exposure to sex hormones had a permanent "organizing" effect on adult mating behavior in the female guinea pig (Phoenix et al. 1959). This study was followed by a number of others, using a variety of animal models of both sexes, that investigated the influence of sex hormones on the brain and on mating behavior. As Bleier (1984: 86) describes it, the theory then

> became generalized not only to behaviors that are more complex than that hormonal regulation of estrous cyclicity, but also to species beyond the rodents who were the original models. It has provided the conceptual framework for theories and beliefs about the sexual differentiation of the human brain and, consequently, of our social roles and behaviors.

Bleier is careful to note that the scientists who used rodent models in their research did not generally extrapolate their findings to humans, but says that, despite their caution, the brain organization framework based on these results has "been uncritically and widely applied by social scientists and natural scientists in other fields" (Bleier 1984: 86). The general story in humans is that androgen exposure shapes the male fetal brain in ways that ultimately account for "masculine" characteristics and behaviors. Females exposed to higher-than-normal levels of androgens, for example, due to congenital adrenal hyperplasia (CAH), are also "masculinized" and are therefore more likely to be tomboys and/or lesbians. In normal female development, lack of androgens results in "feminine" characteristics, such as "passivity, compliance and, bluntly speaking, inferiority" (Bleier 1984: 86). Males whose brains do not receive enough androgen exposure during female development display feminine characteristics and are more likely to be homosexual and/or to fail to develop a clear, male gender identity.

Bleier shows that the simple framework provided by brain organization theory has not been supported by the data. First, even before this research is extended to humans, it does not support a clear developmental trajectory of hormonal brain organization. While the research *did* demonstrate that "hormonal manipulations affected mating behaviors in rodents," it turned out that they did so "often in contradictory and unanticipated ways" (Bleier 1984: 81) For example, administering androgens was found to lead to increased mounting behavior (which was considered to be a male behavior, despite the fact that it occurred in both sexes) in female rats, but the effects were not replicated in female hamsters (Bleier 1984: 106). In general, Bleier argues, these studies are too quick to extrapolate results across species. She cites a study that surveyed the literature to try to determine baseline measurements of the levels of various circulating hormones in different species. The authors determined that "few reports include data for both sexes, for different age groups, or for more than one species" (Bleier 1984: 90). This lack of baseline data is particularly problematic given that different labs measure hormones differently, resulting in variability of measurements that complicates the comparison of studies between, or even within, species.

Bleier also shows that research in humans and other primates failed to show the kind of organizing effects observed in rodents—neither adult hormone cycles nor mating behaviors appear to be affected in ways that support brain organization theory. In rhesus monkey, for example, social rearing conditions have a dramatic effect on making behaviors (Goldfoot et al. 1984). Moreover, subsequent work has shown that the original hypothesis that androgens masculinize the brain was far too simple (a development that has been incorporated into brain organization theory), nor is it possible to distinguish clearly among the effects of estrogens, progestins, and androgens, because all of these groups of hormones are produced by both males and females, and also because there are a multiple different metabolic forms of each hormone, each of which has different physiological effects.

Bleier (1984, 1986) also addresses the issue of sex differences in brain lateralization, pointing to the disagreements among both psychological studies that examine verbal and visuospatial abilities (the abilities most often assigned to the left and the right hemispheres, respectively) and studies that examine sex/gender differences in lateralization. She points out, for example, that the empirical evidence for lateralization differences between women and men is inconsistent. Moreover, the link between the degree of lateralization and performance on specific psychological tasks is tenuous. As noted above, there is some evidence that more boys than girls are mathematically gifted; this has been linked with the claim that the right hemisphere is specialized for visuospatial processing. Bleier, however, claims that it is simply "assumed" that visuospatial processing is "synonymous with mathematical ability" (Bleier 1986: 155). More generally, she claims that the link between greater lateralization in men and their supposedly superior mathematical abilities is based on circular reasoning: "men are superior in visuospatial skills because their right hemispheres are specialized for visuospatial cognitive processing; we know that right hemispheric specialization provides superior visuospatial skills because men have better visuospatial skills than women, who use both hemispheres for visuospatial processing" (Bleier 1986: 154).

6.4 Feminist Empiricism in Science and in Philosophy

Harding (1986, 1993) has famously described work by researchers like Bleier and Fausto-Sterling as "spontaneous feminist empiricism". By this, Harding means that they are scientists, engaging in criticism of other scientists' work, but doing so while sharing with these other scientists a number of basic beliefs about what constitutes good science. According to Harding, spontaneous feminist empiricism is concerned to identify instances in which sexist and androcentric biases prevalent in society influence scientific work (in some cases explicitly and in others without scientists even recognizing their influence). Moreover, this form of feminist empiricist criticism views these biases as "correctable by stricter adherence to the existing methodological norms of scientific inquiry" (Harding 1986: 24). Specifically, it views science influenced by beliefs about sex/gender as "bad science" and, according to Harding, believes that removing the influence of broader social beliefs on scientific research would allow scientists to produce unbiased knowledge. Harding identified the main strength of spontaneous feminist empiricist critiques as strategic: because it focuses on *bad* science, rather than on science in general, it is the kind of feminist critique most likely to be accepted outside of feminist circles. Yet she doubted that these criticisms provided a strong enough foundation to challenge and to change scientific practice. This is because spontaneous feminist empiricism retained the (empiricist) methods and norms of the sciences, which were "too weak to permit researchers *systematically* to identify and eliminate from the results of research those social values, interests, and agendas that are shared by the entire scientific community or virtually all of it" (Harding 1993: 52).

Not long after the "spontaneous" feminist criticisms of authors like Bleier were published, feminist philosophers of science began to develop what Harding called "sophisticated" forms of feminist empiricism. The two most influential of these theories were developed by Lynn Hankinson Nelson and by Helen Longino. Like the spontaneous feminist empiricists, both Nelson and Longino were centrally concerned with sexism in science. But whereas Bleier and other spontaneous feminist empiricists were scientists, Longino and Nelson were coming from a different disciplinary perspective, that of analytic philosophy of science. And regardless of whether Harding is correct to say that spontaneous feminist empiricism viewed (good) science as value-free, this is definitely true of most philosophy of science at the time Harding was writing. Traditionally, philosophers of science have drawn a sharp distinction between the context of discovery and the context of justification. In the former, scientists' idiosyncratic values and beliefs often influenced the hypotheses they proposed. But in the latter context, justifying scientific claims required a rigorous, objective method that ensured that these claims were well-supported, independent of one's values. In other words, good science was value-free. Related to this was the idea that philosophy could study a "rational reconstruction" of a scientific theory, which allowed one to focus on the content of the theory and to ignore the way that the theory was actually developed. Questions about the actual methods and practice of science were the province of historians and sociologists, rather than philosophers (see, e.g., Carnap 1955; Reichenbach 1938).

Nelson and Longino, however, each argued against the idea that (good) science is value-free, and each developed accounts that show the ways in which science requires the influence of values. Yet both were also committed to empiricism, that is, to the view that scientific theories could ultimately be tested only against experience. These two commitments, however, meant that they could not simply condemn sex difference research on the grounds that it reflects sexist (etc.) values. First, since *all* scientific research is shaped by values, philosophical feminist empiricism could not take the avenue of criticism attributed to spontaneous feminist empiricism and simply say that the influence of values on the scientific research made it bad science. Second, the commitment to empiricism meant that it was at least possible that research informed by sexist values was empirically well-grounded. That is, the studies performed on the basis of these assumptions about sex differences may well have been well-designed and conducted and have provided evidence that supported those hypotheses.

In addition to breaking with traditional philosophy of science on the question of values and science, Nelson and Longino also both emphasized the importance of understanding science as a social practice. They therefore rejected the positivist idea that it was sufficient to study the content of a scientific theory (as expressed in a rational reconstruction of a body of scientific knowledge). Instead, both Longino and Nelson developed philosophies of science that examined the ways in which the organization of scientific communities could promote, or fail to promote, good science. It turns out, however, that their focus on scientific communities led both of these philosophers to very different conclusions about brain organization research than were reached by the spontaneous feminist empiricists.

6.5 Nelson and Longino on Brain Organization Theory

Although they both acknowledge the force and the importance of the criticisms of brain organization theory research made by scientists like Bleier, both Longino and Nelson draw much less harsh conclusions about this research, saying that it cannot simply be dismissed as bad science. Longino does acknowledge that some individual studies in this research tradition may be "sloppy, silly, [or] poorly conceived," but claims that "simply to argue against the soundness of any studies that purport to show hormonal determination of sex-linked behaviors" (Longino 1990: 112) does not recognize the complexity of the situation. Similarly, Nelson disagrees with the feminist critics who conclude that research on "sex differentiated lateralization was never warranted by available evidence, but motivated solely by social and political interests" (Nelson 1995: 411).

Instead, both argue that research on the influence of sex hormones in the brain has to be understood in the context of the scientific communities by which it was conducted. Only after dissenting (including feminist) viewpoints were available did the shortcomings of the research become clear. As Longino puts it, "the very existence of controversy [...] makes the task of analyzing these studies much more feasible than it would be in the absence of controversy" (Longino 1990: 117).

Yet it is important to note that the controversy to which Longino refers is, in her view, primarily due to changes in the sociopolitical climate, rather than to disagreements that focus on the evidence. Only the development of alternative viewpoints allowed the evidence to be reconsidered, a point to which I will return later. For Longino, the key issue to be addressed is *how* this broader climate influenced the kind of studies conducted and the kind of conclusions accepted by a scientific community. This way of describing the relevant questions about this research draws on Longino's distinction between "constitutive" and "contextual" values. The former are "generated from an understanding of the goals of science" and "determin[e] what constitutes acceptable scientific practice or scientific method." The latter are "personal, social, and cultural values, those group or individual preferences about what ought to be" and "belong to the social and cultural environment in which science is done" (Longino 1990: 4). Contra the traditional view that the latter kind of value should play no role in science, Longino argues that they not only can, but *must* do so. Thus, in her analysis of brain organization theory, her aim is to show, first, how contextual values play a role "in assigning evidential relevance to the data" (Longino 1990: 113) and, second, how they "motivate the acceptance of global framework like assumptions that determine the character of research in an entire field" (Longino 1990: 86).

Longino's assessment of the way that contextual values shaped research on hormones and sex/gender differences is complex. Unlike Bleier, she does not view the sex- or gender-related values that shape this research as expressing either sexism or androcentrism, but she does point out that the studies embody "a third expression of patriarchal values, the assumption of thoroughgoing dimorphism or sexual essentialism" (Longino 1990: 129). Yet, again, this problematic assumption does not make the research bad science. Rather, for Longino (1990: 162, emphasis added), "the label 'bad science' can only be applied in light of criteria that are operative *within* a particular field of research". Similarly, she points out that "many cases of good science, or of scientific inquiry that conforms to the constitutive values of a particular tradition, can be shown to be bad science if considered out of the larger research contexts that give them shape and meaning" (Longino 1990: 112).

As Clough (1998, 2003) has argued, by making these claims Longino relativizes what counts as evidence to a particular framework or set of constitutive values and thus shifts the debate from one of whether there is good evidence for a claim to one of whether the contextual values that comprise a particular framework can be justified. Key here is the point that, for Longino (1990: 75), interpretive frameworks and background assumptions "may not be subject to empirical confirmation or disconfirmation, and [...] may be infused with metaphysical or normative considerations." This means that critically assessing them is not a matter of simply determining whether they are supported by the available evidence.

At first glance, Nelson's version of a feminist empiricism appears to provide better resources for criticizing problematic research. Whereas Longino distinguishes between the empirical evidence supporting a theory and the background assumptions and contextual values that make that evidence plausible, Nelson (following Quine) takes a holistic view on which *all* of these things are evidence: "holism is best understood as taking the *evidence* for a specific claim or theory to include *both* the observational consequences of that claim or theory (together with the larger theory or theories within which it is embedded) *and* the relationship of the claim or theory in question to other current theories, methods, standards, and practices" (Nelson 1995: 405). This is in contrast with views that construe evidence "narrowly (e.g., as exhausted by 'data' or observation sentences), relegating other theories and broad methodological and metaphysical assumptions figuring in the adjudication of a specific theory to a realm of 'background' assumptions, 'auxiliary' theories, or 'disciplinary matrices'" (Nelson 1995: 405). Whereas Longino distinguishes between empirical evidence and the framework or criteria that organize this evidence according to a particular research tradition, for Nelson, all of these considerations are evidence and can thus be brought to bear on the assessment of a theory. Moreover Nelson's (1995: 405–6) empiricism "leaves intact the norms of empirical success, explanatory power, and predictive success", thus giving clear standards for this assessment.

Yet because Nelson, like Longino, views the scientific community as the ultimate arbiter of what counts as scientific knowledge, she also ends up relativizing knowledge to the communities that produce it. A significant departure of Nelson's theory from Quine's is that Nelson's holism is limited. While Quine thought that the whole body of our knowledge was tested with any observation, Nelson (1995: 418) says instead that "[i]n neither science nor the philosophy of science is the whole of a current theory of nature drawn on in the adjudication of particular claim and theories." Thus while "we experience the world through the lens of our going theories" (Nelson 1990: 112), any of the various elements of the theory can be tested by new data. But they need not be so tested. Nor does Nelson offer guidance about *when* different aspects of our going theories (including social and political values) should be tested. Rather, it is up to individual communities, which set their own standards: "there is no algorithm [...] for demarcating the theories and standards, communities or experimental results which should figure in the evaluation of a theory and/or of its adoption" (Nelson 1995: 417).[1]

These latter aspects of Nelson's theory lead her to draw much weaker conclusions about brain organization theory than did the spontaneous feminist empiricists. In particular, Nelson provides an extended case study of Geschwind and Behan's (1982) hypothesis that, because testosterone slows the development of the left hemisphere, men are right-hemisphere dominant. This hypothesis was originally

[1]Nelson (1995: 418) does, however, acknowledge that "scientists and philosophers of science may appropriately consider theories and results which those involved did not when evaluating a research hypothesis or theory".

based in part on a study in rats that showed that two areas of the cortex were thicker in the right hemisphere than in the left, in male, but not female, animals. The rationale for extending this hypothesis to humans was a study in humans that showed that two areas of the right hemisphere develop several weeks earlier than the corresponding areas in the left hemisphere.

Although she acknowledges the arguments of authors like Bleier and Fausto-Sterling that this research was not well-justified by the available evidence, Nelson herself concludes that this is only true of the Geschwind and Behan hypothesis when it is considered in isolation. On her own holistic account, however, the evidence for the hypothesis includes "both direct empirical success (the successful prediction and explanation of relevant data) *and* how the hypothesis or program is integrated into other accepted theories and research" (Nelson 1995: 411). In fact, she argues that, at least until the early 1980s, there *was* evidence supporting the Geschwind-Behan hypothesis, evidence that included cultural assumptions about cognition and gender, as well as scientific research in several related fields. The scientific evidence was informed by the framework provided by brain organization theory, including the early studies in reproductive endocrinology that showed that testosterone had an organizing effect on the fetal rat brain, which determined hormone cycles in the adult, as well as later studies, also mainly in rats, that extend the theory to say that hormone exposure also affected other spheres of sex-differentiated behavior (e.g. aggression and maze-running performance). There was also then-recent work in songbirds linking testosterone exposure, brain morphology, and the ability to sing (Nottebohm and Arnold 1976). In addition, there was evidence from psychology that claimed to establish the existence of sex differences in, e.g. mathematical abilities, and of the lateralization of cognitive abilities. Finally, the correlations among left-handedness, immune system disorders, and learning disabilities, as well as the greater prevalence of all of these characteristics in males, had been established prior to Geschwind and Behan's (1982) paper; the contribution of these authors was to bring together these separate lines of investigation and hypothesize that testosterone slowed left hemisphere development.

Nelson (1995: 413) thus concludes:

> When evidence is construed holistically, Geschwind and Behan's hypothesis is revealed to be neither far fetched, nor purely politically motivated. It represented the synthesis of central research questions, current hypotheses, and experimental results in three research traditions, and the convergence of these with cultural assumptions about sex differences (e.g. that males have superior spatial and mathematical abilities, that there is a biological foundation for sex differences, etc.).

Two aspects of Nelson's and Longino's conclusions about brain organization theory are striking. First, they appear to be ignoring or excusing some of the very real problems with this research. I think that neither Nelson nor Longino give sufficient weight to the empirical criticisms levied by feminists against brain organization theory. For example, the study by Chi et al. that was used as evidence for the Geschwind and Behan hypothesis showed that parts of the left hemisphere develop more slowly than the corresponding areas on the right, but they *didn't* show that any sex differences in this pattern. Moreover, while there was evidence for sex differences in brain lateralization, Bleier (1984: 92–3) points out that there was empirical disagreement among researchers:

> One research group finds that women are *less* lateralized than men [...] Another group finds that women are *more* lateralized [...] A third group finds no convincing sex differences in cognitive abilities or hemispheric lateralization, and yet a forth [...] questions the basic assumption that there is any lateralization of cognitive functions at all, quite apart from the issue of sex differences.

These are serious empirical problems, regardless of whether one takes a feminist perspective on this research or not. Yet discussions of projects like Bleier's by feminist philosophers tend to focus mainly on the *reason* that research with such serious empirical flaws was so widely accepted; i.e., that it reflected prevailing beliefs about gender and gender differences. Bleier *does* draw this conclusion, but she does so in an attempt to explain the results of her empirical analysis. These two aspects of her project are separate, in that her empirical criticisms stand even without the additional identification of the role of sexist beliefs in scientists' acceptance of brain organization theory. Feminist philosophers, however, have generally focused mainly on the social aspects of this critique.

This focus on social aspects is also relevant to the second problem I identified with Longino and Nelson, which is the tension between their claim that brain organization research is not bad science and their claim that the science would have been better if it took feminist criticisms into account. For example, Nelson (1995: 417) says that, despite not being bad science, "the assumption of a hormonal basis for sex-differentiated lateralization and the [Geschwind-Behan hypothesis] are revealed to be substantially *less warranted* than when judged without the benefit of these more exacting critiques." This is something that she and Longino have in common; they both believe that prior to the availability of feminist criticisms, the research community engaging in various aspects of brain organization theory cannot be faulted. In the context within which they were working, the hypotheses tested and the conclusions accepted were indeed justified. The addition of feminist criticisms to the available evidence, broadly construed (in Nelson's theory) or the challenges that feminists raised against patriarchal background assumptions (in Longino's) changed the nature and composition of the community relevant to the assessment of brain organization research. As Nelson (1995: 414) puts it: "[t]he recognition of the role of androcentrism and assumptions about gender in the original research...attests to the *integration* of results, standards, theories, and research backgrounds of scientific disciplines and *feminist* communities." Similarly, one of the major claims of Longino's epistemology is that the inclusion of diverse viewpoints in a scientific community is central to the community's ability to produce objective knowledge.

6.6 Feminist Philosophy of Science and the Turn to Science Communities

The second striking aspect of Longino's and Nelson's criticisms is that both of them want to claim that a research program that took the criticisms of these feminist scientists into account would be *better* than one that did not, but also to avoid criticizing the research that was conducted as bad science. Instead of attempting to provide guidelines for assessing empirical evidence or theories, both Longino and Nelson turn to the question of what sorts of *communities* produce the best knowledge. Moreover, because each of these philosophers was centrally concerned with showing that values play a role even in *good* science, their characterization of "good science" tended to center on objectivity; they wanted to show that science can still be objective though influenced by values. Thus, Longino (1990, 2002) characterizes an ideal epistemic community in which objectivity is maximized through the inclusion of diverse viewpoints, and she provides a set of norms that serve to promote objectivity. Nelson focuses on the specific social processes that characterize science, claiming that the normative questions that occupy philosophers of science should address whether such scientific practices as "recruitment or education of scientists [...] peer review mechanisms [etc.] are the processes that *should* be at work—are likely to produce the *best* theories and research programs" (Nelson 1997: 103).

This focus on communities and their standards has been criticized by other feminist philosophers of science. Of particular importance is Clough's (1998) objection that the turn to the analysis of epistemic communities amounts to "a hasty retreat from evidence". Clough argues that feminist philosophers who claim that data can only be assessed as potential evidence for/against a theory relative to an interpretive framework (or a set of background assumptions or community standards) invite a skepticism that threatens to undermine feminist criticisms of science.[2] Longino and Nelson, for example, draw on the underdetermination thesis to argue that, because it is always possible for more than one theory to accommodate a data set (or any possible set of data), the choice among rival theories always depends in part on background assumptions or interpretive frameworks that can include social and political values. But without the ability to appeal to shared evidentiary standards, there is no good empirical way to decide between frameworks. I have argued above that this is indeed the case for feminist criticisms of brain organization theory; Longino's and Nelson's discussions of this research back away from strongly critical conclusions offered by the spontaneous feminist empiricists.

---

[2]Clough's (1998) paper focuses on Longino's work, as well as that of Sandra Harding and Evelyn Fox Keller; in a 2004 paper, however, she argues that Nelson's Quinean approach runs the same risks. Similarly, Maya Goldenberg, whose work I will discuss shortly, shows that Clough's criticisms also apply to Nelson's feminist empiricism. Interestingly, Clough (2004) also identifies a split between evidence and framework in Bleier's writing.

Perhaps this explains, too, why more recent criticisms of sex/gender difference research (e.g., Kaiser et al. 2009; Fine 2010; Jordan-Young 2010; Meynell 2012; Bluhm 2013a, b) do not tend to engage with feminist philosophy of science. Instead of drawing on the theoretical resources provided by feminist empiricism, these critics are engaging in much the same kind of critique practiced by Bleier and the other spontaneous feminist empiricists. The relativism of Longino's and Nelson's conclusions is not very satisfying, given the very real problems with this science. Moreover, as I will show below, the turn to analyzing science communities does not allow us to detect some of the important roles that values play in science. Yet despite this, I believe that feminist philosophy of science *does* have something to offer feminist critics of science. In making this case, I will draw on a recent paper by Goldenberg (2015), in which she looks at the potential implications of feminist empiricism for debates in a different scientific context, that of clinical decision-making.

Goldenberg distinguishes between two streams in feminist philosophy of science. The dominant stream focuses on "community-based social knowledge;" it adopts the basic tenets that "(1) knowledge production is a social process and (2) communities rather than individuals are the agents of knowledge" (Goldenberg 2015: 5). Nelson and Longino are the two leading exemplars of this tradition. The less-developed stream of feminist empiricism, Goldenberg calls the "values as evidence" stream. As the name suggests, feminist philosophers of science who take this position view the sorts of contextual values that make up the interpretive frameworks of the "community-based social knowledge" stream as being themselves subject to empirical testing. Clough, with her denial of the idea that value-laden interpretive frameworks are entirely distinct from evidence, is one of two authors that Goldenberg characterizes as belonging to this stream. The other is Elizabeth Anderson (2004), who has argued that the choice between two value-laden frameworks can indeed be determined on the basis of empirical considerations (both the empirical evidence and such traditional empiricist criteria as the fruitfulness of two alternative research approaches).

Both of these streams are centrally concerned with the role of values in science. I have already described Clough's criticisms of the "community-based social knowledge" stream; by relativizing evidence to a particular framework, it undercuts feminist criticisms of science. Goldenberg further shows that this steam of feminist empiricism is concerned with characterizing *idealized* epistemic communities, ones in which the operative social processes are the right ones to produce good, objective science and to manage diverse values. This approach, however, "establishes good evidence through the diversity of the contextual values represented and not the *content* of those value judgments" (Goldenberg 2015: 14).

Goldenberg points out that these idealized accounts are not of much help in the case of real scientific communities. She uses the case of biomedical research, influenced by evidence-based medicine (EBM), to show that critics of EBM often draw on the same themes in philosophy of science that concern feminist philosophers (e.g., underdetermination) and argues that only the values as evidence stream offers critical resources that allow physicians to make good clinical decisions. The community-based social knowledge stream does not help because the "current state of the biomedical community of knowers is a far cry from the feminist ideal community" (Goldenberg 2015: 21).

By contrast, the values as evidence stream provides resources for negotiating a critical analysis of the evidence produced by real epistemic communities that fall short of the ideal. Goldenberg provides an example of a physician who must assess the value of a clinical trial of a promising new drug in determining the best treatment for a patient. While the trial shows that the drug is superior to standard therapy, she realizes that in the study the standard treatment was given at a lower dose than would normally be used. Since she *also* sees that the first author has received grants from the pharmaceutical company that produces the drug, and that the company sponsored the trial, she chooses not to suggest the new therapy to her patient, and to recommend the standard treatment instead. Crucially, the physician's assessment is not based solely on social or political considerations, nor solely on evidence, but on the way that they are intertwined.

Goldenberg's distinction is helpful for understanding why Nelson and Longino do not give adequate weight to Bleier's criticisms of brain organization theory. Their philosophical approach focuses on science communities and the social context within which those communities conduct their research and, because of this, tends to address the roles that values play at a broader, social level. This approach also suggests that the best way to change scientific theories is to change the scientific community; both Nelson and Longino therefore focus on developing appropriate norms for the composition of scientific communities, aiming to increase the objectivity of the knowledge they produce (e.g., Longino 1990, 2002; Nelson 1990, 1993). The goal of this approach is best expressed by Sandra Harding, who believed that the careful empirical criticisms made by the spontaneous feminist empiricists did not provide sufficient resources for changing sexist science. Instead, she noted that, historically, "[i]t is movements for social liberation that have most increased the objectivity of science, not the norms of science as they have in fact been practiced, or as philosophers have rationally reconstructed them" (Harding 1986: 25).[3]

The community-based social knowledge approach has failed in the case of brain organization theory. Many important social and political changes have occurred in the past 50 years that have challenged the values that informed the early research. In particular, ideas about gender roles and gender-appropriate characteristics have shifted dramatically, and the very idea that masculinity and femininity are biologically determined has been seriously questioned. The overt sexism that characterized many discussions of gendered behavior is no longer generally accepted. Yet, this has not led to the kind of revision of values and beliefs envisaged by feminist philosophers. Instead of being rejected, if anything, brain organization theory has become more firmly entrenched than ever. Jordan-Young (2010: 37)

---

[3]Harding herself viewed feminist standpoint theory as the best way of developing such frameworks, though she did consider it possible that the "sophisticated" feminist empiricisms developed by Longino and by Nelson could also do so.

notes that in recent studies, "the idea of hormonal organization of the brain is now only rarely identified as a theory and is usually incorporated into research as a 'background fact' of development." Nor have the feminist criticisms of brain organization theory been taken up by scientists working within the framework. In a paper marking the 50th anniversary of the Phoenix et al. paper that first identified the organizing effects of prenatal testosterone exposure in the female guinea pig, Arnold (2009: 570) says that this experiment "provided a conceptual framework that has been repeatedly tested and improved since 1959, but has not been substantially undermined by experimental findings in the intervening half century." The addition of feminist views to science has not led to the improvements in science that Nelson and Longino believed would occur.

6.7 Values and Evidence in Brain Organization Theory

Instead, I believe, feminist philosophy of science must draw on and further develop the insights of the values as evidence stream in order to adequately understand and criticize brain organization theory. While the community-based social knowledge stream has done important work in identifying and recognizing the importance of the social dimensions of science, it has been of less help in developing feminist criticisms of specific research programs. What is required now is careful empirical analysis that is sensitive to the complex relationship between science and society, and between values and evidence. Recall that Goldenberg characterizes the dominant (community-based social knowledge) stream of feminist empiricism of paying insufficient attention to the *content* of the values that inform scientific research. Clough (2003: 27–28) argues that, instead of examining the role of broad social frameworks, "[m]oving toward more local, empirical examinations of science and other social institutions might allow for more sensitive and fine-grained analyses of the roles of sex/gender, race, and class." I believe that it is necessary to develop fine-grained analyses that attend carefully to the content of gender-related values. This is because, in addition to their broad effects on the assumptions made or background beliefs held regarding gender, social and political values also permeate empirical inquiry, influencing the details of the experiments conducted and the data gathered.

Goldenberg does credit Nelson and Longino with recognizing that evidence is value-laden, but says that their approach does not adequately recognize the extent to which the evidence can support one set of values over another (I suspect that this is why, despite claiming that the feminist criticisms of brain organization theory can lead to better science, they do not want to say that brain organization theory is not well-supported by the evidence.). To this, I would add that the kind of close analysis of the evidence that Clough advocates permits a more nuanced understanding of the way that values are relevant to, and reflected in, empirical evidence. In supporting this claim, I will draw on two issues raised in Jordan-Young's analysis of brain organization theory.

In her recent book, Jordan-Young carefully and systematically analyses nearly 25 years of studies on human beings that fall under the framework provided by brain organization theory. Much like Bleier, she concludes that "the evidence simply does not support the theory" (Jordan-Young 2010: xiii). Although Jordan-Young herself largely refrains from discussing the broader social and political implications of her empirical critique (see Jordan-Young 2010: 3), it provides an important resource for an explicitly feminist analysis of the scientific research. It also shows that, contra the predictions of the community-based social knowledge stream, broad social changes in our understanding of gender have actually made brain organization theory *more* problematic from a feminist perspective. Briefly, as these changes in broader social beliefs about gender and about sex/gender differences have occurred, they become reflected in the *kinds* of differences that researchers measure. While it is true that broader social changes mean that scientists are working with a less overtly sexist set of beliefs about the differences between women and men, they still accept the kind of thoroughgoing sex essentialism identified by Longino. In other words, men and women are still viewed as fundamentally different, though the specific differences thought to exist have changed. Jordan-Young's assessment of the empirical evidence for brain organization theory makes it clear that the basic framework it provides (i.e. that women and men are different because their brains are shaped by different levels of sex hormones during fetal development) is abstract enough that it can be filled in with very different stories about the nature of these differences. As a result, brain organization theory appears to have much more support than it actually does.

The first example from Jordan-Young's book deals with the way that brain organization theory has thought about men's and women's sexuality and sexual behaviors. She says that during her research, she often heard that it was simply common sense that male sexuality and female sexuality are different. This belief functions as a background "framework" value. At a more fine-grained level, however, things are more complicated. Jordan-Young identifies two distinct periods of research, during which the differences between masculine and feminine sexuality are viewed very differently. From the late 1960s up to the beginning of the 1980s, "brain organization researchers relied on a model of human sexuality that sharply divided masculine and feminine sexual natures" (Jordan-Young 2010: 113). Men were initiators of sexual contact, were versatile in their behaviors (e.g., enjoyed a number of different sexual positions), and their sexuality was genitally focused. They also masturbated, desired multiple sex partners, and could be interested in sex without any romantic attachment. By contrast, women were "receptive," inclined to monogamy and romance, did not masturbate, and were "conservative" with regard the kinds of sexual activities and positions they enjoyed. For both men and women, attraction solely to members of the opposite sex was central to normal sexuality. This early model informed studies of women with congenital adrenal hyperplasia (CAH), who had been exposed to high levels of testosterone during fetal development and who were therefore expected to have characteristically masculine sexuality. Researchers compared the occurrence of (self-reported) "male" sexual behaviors by these women, often comparing the reports to those of women without CAH.

But, starting in the early 1980s, the model of sexuality that was used by researchers began to change (Jordan-Young points out that even by the early 1970s, there was already a lot of evidence to show that "normal" women engaged in male sexual behaviors.). First, since this time "the great majority of studies that include *any* information about sexuality have exclusively focused on sexual orientation, omitting mention of any other aspects, such as libido, arousal patterns, and so on" (Jordan-Young 2010: 131). Moreover, behaviors that had once been "unequivocally coded as masculine" (e.g., masturbation, having multiple partners, having sex outside of marriage) "came to be understood as 'commonsense' features of feminine sexuality" (Jordan-Young 2010: 113–14), undermining the results of the earlier studies in women with CAH.

Jordan-Young draws out the implications of this shifting understanding of masculine versus feminine sexuality for brain organization theory, making it clear that studies that make completely incompatible assumptions about sexual behavior are lumped together as evidence for the theory. Her close analysis shows that the evidence does not support brain organization theory and, key for the purposes of this chapter, it also shows how complex a role gender-related values play in science. As society moved away from the sexist conception of women as passive and focused on monogamous, romantic relationships (and of men as active and as interested in sex for its own sake), the belief persisted that feminine and masculine sexuality are distinct. At the level of specific scientific studies, these beliefs interacted to influence exactly what was measured and what behaviors counted as evidence for brain organization theory.

A second way in which values and evidence are intertwined in this area of research is illustrated by a subtle shift in the explanation of why women's and men's cognitive and behavioral characteristics differ. Jordan-Young points out that scientists tend not to talk about sex/gender differences in *ability*, but rather describe gendered traits in terms of the different *interests* of each sex. This allows scientists to avoid charges of sexism, since "on a gut level, interests seem to be more value-neutral than skills" (Jordan-Young 2010: 202). Sex-typed interests are seen as "both preceding and amplifying other psychological differences, such as those in specific cognitive skills" (Jordan-Young 2010: 200). Thus, the reason that there are not a lot of women engineers is that women are just not interested in engineering. It is not that they naturally lack the necessary mathematical ability and visuospatial skills; it's that they choose not to develop them.

Yet, while Jordan-Young does not address this point, it's not that long ago that scientists were quite comfortable talking about natural differences in ability. An article in *Science* describing Benbow and Stanley's research on the greater prevalence of high mathematical achievement in boys than in girls has the headline "Math Genius May Have Hormonal Basis" (Kolata 1983). The original article concludes that the "[r]easons for this sex difference are unclear" (Benbow and Stanley 1983), but the authors have just finished explaining why various environmental factors fail to explain their data, and in a footnote to this conclusion, they cite a number of "possible endogenous influences." It is clear that they favor a biological explanation of these performance differences.

Moreover, citing "interests" rather than "ability" as the reason for sex differences in achievement, particularly in studies that otherwise draw on brain organization theory, is completely compatible with biological determinism. Recall Doell and Longino's (1988) description of the linear model of development, which traces sex differences from chromosomes to hormones, to brains, to behavior. Adding "interests" between "brains" and "behavior" does not make resulting sex differences any less determined by hormone exposure (see also Jordan-Young 2010: 198–201). If women are doomed to have the stereotypical female interests of "nurturing and self-decoration" and males' interests naturally "revolve around action, social domination, and mastery of skills" (Jordan-Young 2010: 203), it seems that the perceived "value neutrality" of focusing on interests just masks the same old sexism.[4]

These two examples show that the basic (problematic) framework of brain organization theory persists despite important changes in social views about sex/gender differences. This is because the theory is abstract enough to incorporate very different ways of understanding and explaining these differences. To understand how this occurs, we need to develop an analytic approach that brings together traditional feminist concerns about gender and society with careful empirical analysis that can detect the way these social views are reflected in scientific research. Because neuroscientists continue to search for biological explanations of the purported cognitive and behavioral differences between women and men, and because this search continues to owe more to broadly accepted beliefs about gender than to evidential support, feminist philosophy of science must continue to develop tools that can detect the subtle ways that values interact with empirical research.

References

Anderson, E. (2004). Uses of value judgments in science: A general argument, with lessons from a case study of feminist research on divorce. *Hypatia, 19*(1), 1–24.

Arnold, A. P. (2009). The organizational-activational hypothesis as the foundation for a unified theory of sexual differentiation of all mammalian tissues. *Hormones and Behavior, 55*, 570–578.

Benbow, C. P., & Stanley, J. C. (1983). Sex differences in mathematical reasoning ability: More facts. *Science, 222*(4627), 1029–1031.

Birke, L. (1986). *Women, feminism and biology: The feminist challenge*. Brighton: Harvester Press.

Bleier, R. (1984). *Science and gender: A critique of biology and its theories on women*. New York: Pergamon Press.

Bleier, R. (1986). Sex differences research: Science or belief? In R. Bleier (Ed.), *Feminist approaches to science* (pp. 230–245). New York: Pergamon Press.

[4]Similarly, I have argued that current pop neuroscience books tend to give biological explanations for sex/gender differences in personality characteristics and preferences, in contrast with earlier work that viewed gender *roles* as biologically determined. This shift does not, however, prevent the current work from arguing that the social status quo is inevitable (Bluhm 2012).

Bluhm, R. (2012). Beyond neurosexism: Is it possible to defend the female brain? In R. Bluhm, H. Maibom, & A. J. Jacobson (Eds.), *Neurofeminism: Issues at the intersection of feminist theory and cognitive neuroscience* (pp. 11–29). Hounsmills, Basingstoke: Palgrave.
Bluhm, R. (2013a). New research, old problems: Methodological and ethical issues in fMRI research examining sex/gender differences in emotion processing. *Neuroethics, 6*(2), 319–330.
Bluhm, R. (2013b). Self-fulfilling prophecies: The influence of gender stereotypes on functional neuroimaging research on emotion. *Hypatia, 28*(4), 870–886.
Carnap, R. (1955). Logical foundations of the unity of science. In O. Neurath, R. Carnap, & C. Morris (Eds.), *International encyclopedia of unified science* (Vol. 1, pp. 42–62). Chicago, IL: University of Chicago Press.
Clough, S. (1998). A hasty retreat from evidence: The recalcitrance of relativism in feminist epistemology. *Hypatia, 13*(4), 88–110.
Clough, S. (2003). *Beyond epistemology: A pragmatist approach to feminist science studies*. Lanham, MD: Rowman and Littlefield.
Clough, S. (2004). Having it all: Naturalized normativity in feminist science studies. *Hypatia, 19* (1), 102–118.
Doell, R. G., & Longino, H. E. (1988). Sex hormones and human behavior: A critique of the linear model. *Journal of Homosexuality, 15*(3–4), 55–78.
Fausto-Sterling, A. (1985). *Myths of gender: Biological theories about women and men*. New York: Basic Books.
Fine, C. (2010). *Delusions of gender: How our minds, society, and neurosexism create difference*. New York: W.W. Norton & Company.
Geschwind, N., & Behan, P. (1982). Left-handedness: Association with immune disease, migraine, and developmental learning disorder. *Proceedings of the National Academy of Sciences, 79*, 5097–5100.
Geschwind, N., & Galaburda, A.M. (1987). *Cerebral lateralization: Biological mechanisms, associations, and pathology*. Cambridge MA: MIT Press.
Goldenberg, M. J. (2015). How can feminist theories of evidence assist clinical reasoning and decision-making? *Social Epistemology*, *29*(1), 3–30.
Goldfoot, D., Wallen, K., Neff, D. A., McBrair, M. C., & Goy, R. W. (1984). Social influences on the display of sexually dimorphic behavior in rhesus monkeys: Isosexual rearing. *Archives of Sexual Behavior, 13*(5), 395–412.
Harding, S. (1986). *The science question in feminism*. Ithaca, NY: Cornell University Press.
Harding, S. (1993). Rethinking standpoint epistemology: "What is strong objectivity?". In E. Potter & L. Alcoff (Eds.), *Feminist epistemologies*. London: Routledge.
Harris, G.W., & Levine, S. (1965). Sexual differentiation and its experimental control. *Journal of Physiology, 181*(2), 279–400.
Jordan-Young, R. M. (2010). *Brain storm: The flaws in the science of sex difference*. Cambridge: Harvard University Press.
Kaiser, A., Kuenzli, E., Zappatore, D., & Nitsch, C. (2009). On sex/gender related similarities and differences in fMRI language research. *Brain Research Reviews, 61*, 49–59.
Kolata, G. (1983). Math genius may have hormonal basis. *Science, 222*, 1312.
Longino, H. E. (1990). *Science as social knowledge*. Princeton: Princeton University Press.
Longino, H. E. (2002). *The fate of knowledge*. Princeton: Princeton University Press.
Meynell, L. (2012). The politics of pictured reality: Locating the object from nowhere in fMRI. In R. Bluhm, H. Maibom, & A. J. Jacobson (Eds.), *Neurofeminism: Issues at the intersection of feminist theory and cognitive neuroscience* (pp. 11–29). Hounsmills, Basingstoke: Palgrave.
Nelson, L. H. (1990). *Who knows? From Quine to a feminist empiricism*. Philadelphia: Temple University Press.
Nelson, L. H. (1993). A question of evidence. *Hypatia, 8*(2), 172–189.
Nelson, L. H. (1995). A feminist naturalized philosophy of science. *Synthese, 104*, 399–421.
Nelson, L. H. (1997). Empiricism without dogmas. In L. H. Nelson & J. Nelson (Eds.), *Feminism, science, and the philosophy of science* (pp. 95–199). Dordrecht: Kluwer.

Nottebohm, F., & Arnold, A. (1976). Sexual dimorphism in vocal control areas of the songbird brain. *Science, 194*, 211–213.

Phoenix, C. H., Goy, R. W., Gerall, A. A., & Young, W. C. (1959). Organizing action of prenatally administered testosterone propionate on the tissues mediating mating behavior in the female guinea pig. *Endocrinology, 65*, 369–382.

Reichenbach, H. (1938). *Experience and Prediction: An analysis of the foundations and the structure of knowledge*. Chicago: University of Chicago Press.

Chapter 7
The Reason/Emotion Divide in Contemporary Philosophy of Psychology

Michelle Maiese

7.1 Introduction

Reason and emotion have long been regarded as separate faculties or functions of the mind, and customary conceptualizations of them have incorporated numerous dichotomies: active versus passive, intellectual versus bodily, objective versus subjective, and reflective versus intuitive. These dichotomies, in turn, have been associated with binary constructions of gender difference. Indeed, throughout the history of Philosophy, men have tended to be viewed as more active, intellectual, and rational, whereas women have tended to be viewed as more passive, corporeal, and emotional. Furthermore, because this tradition has associated emotion primarily with irrationality, subjectivity, and chaos (Lutz 1986), women have been viewed not only as naturally more emotional, but also as naturally inferior. Thus, the alleged distinctions between emotion and thought, between affect and cognition, and between passion and reason have served to reinforce the notions that women and men are fundamentally different, and that men are better thinkers. Moreover, this assumption that reason is separate from emotion cannot be brushed aside merely as an outdated way of thinking since it continues to shape philosophical discourse. This is evidenced, I will argue, by much of the work being done in contemporary Philosophy of Psychology. Although theorists writing in this field seldom comment directly on the reason/emotion divide, let alone discuss how their research relates to issues of gender, much of this work incorporates the assumption that reason and emotion are somehow distinct faculties and presumes that reason and abstract theorizing (rather than emotion) assume center stage in sophisticated cognitive functioning. I will maintain that if we reject this reason/emotion divide,

M. Maiese (✉)
Emmanuel College, Boston, MA, USA
e-mail: maiesemi@emmanuel.edu

M.C. Amoretti and N. Vassallo (eds.), *Meta-Philosophical Reflection on Feminist Philosophies of Science*, Boston Studies in the Philosophy and History of Science 317, DOI 10.1007/978-3-319-26348-9_7

we thereby can undermine any ideological support it provides in favor of dichotomous gender categories. In addition, we likely will arrive at a more accurate picture of cognitive functioning.

7.2 The Reason/Emotion Divide in the History of Thought

Lloyd's well-known work describes how "our ideals of Reason have historically incorporated an exclusion of the feminine" and how "femininity itself has been partly constituted through such processes of exclusion (Lloyd 1984: x). Men typically have been associated with the intellect and activity, whereas women have been associated with bodily perturbations and passivity. For example, articulations of the form-matter distinction among the early Greeks align maleness with active, determinate form, and align femaleness with passive, indeterminate matter (Lloyd 1984: 3). Likewise, the deeply entrenched notion that emotion is opposed to reason can be traced to Plato's doctrine of the tripartite soul in the *Republic* and *Phaedrus*, where we learn of the supposed dichotomy between the masterly "reasoning" part of the soul on the one hand, and the subservient "spirited" and "appetitive" (or passionate) parts of the soul on the other. This early work posits a dichotomy between the intellect and the body and characterizes intellectual life as a matter of purging the rational soul from the "gross intrusions" of the body (Lloyd 1984: 6). Plato likens reason to a charioteer whose job it is to guide emotion and desire and keep these two unruly "horses" from getting out of control and leading us astray. According to this view, emotions are inherently disruptive and overwhelming forces not under our direct control, and they involve intrinsically passive feelings that are at best arational, and at worst downright irrational. In short, this dichotomous picture tells us both that our rationality is inherently *non*-emotional, and also that our emotions are inherently *non*-rational. To live virtuously, a subject's capacity for reason must gain control over the lower aspects of human nature, such as sense perception and bodily passion. This often is construed, Lloyd (1984: 26) maintains, as a matter of "shedding the influence and intrusion of femaleness" and becoming male. Aristotle, in particular, is infamous for his suggestion that women follow their passions more readily, that their powers of reason are unstable, and that this renders them less capable of philosophical contemplation and intellectual pursuits.

A slightly more nuanced view can be found in the work of Rousseau. While Plato emphasizes the need for rational control, and theorists such as Aristotle present femininity as "somehow derivative in relation to a male paradigm of rational excellence" (Lloyd 1984: 37), Rousseau views the minds of women as different in a way that makes them "complementary" to the minds of men (Lloyd 1984: 75). While it is inappropriate for them to engage in male-style intellectual pursuits (i.e. the search for abstract truths), women possess traits such as taste, practical sense, and feeling that make them well-suited to promote human well-being within the domestic domain. According to Rousseau, the home ought to be a haven of virtue and tenderness, and men ought to be "nurtured in childhood" and "tended in manhood" by virtuous women (Lloyd 1984: 78). However, he thinks it is important to keep women out of the public sphere since feeling and emotion pose a threat to the proper functioning of the State. Because passion involves subjective bias and often leads to poor judgments or erratic behavior, the story goes, emotion can be understood as a disruption to the sort of rational thinking needed to guide public affairs.

It seems clear that throughout the history of Philosophy, men have been viewed as more rational and therefore better suited for the world of work and politics, whereas emotion and the bodily dimension of human existence have been associated with women. And philosophical texts are not alone in expressing such views. As Lutz (2001: 104) notes, there is a long history of clinicians who have associated emotionality with normal female functioning as well as with deviance. Pathology often has been emotionalized (e.g., in longstanding conceptions of female hysteria), whereas cool-headed reasoning continues to be viewed as masculine. Likewise, Mumby and Putnam maintain that the concept of "bounded rationality" that assumes center stage in classical organizational theory "stems from a system of binary opposites that privileges the masculine over the feminine" (Mumby and Putnam 1992: 468). Here, rationality often is construed as a matter of "satisficing," means-ends reasoning, and incremental decisions, and cognition (rather than emotion) remains the central link between thought, action, and choice. Holistic forms of reasoning such as intuition, on the other hand, are treated as nonrational, and decisions based on emotions are viewed as irrational (Simon 1989). This is related to the fact that organizational practices construct the identities of men and women very differently, with women often cast as marginalized actors who contribute only to the provision of support services, nurturance, and human relations within the workplace (Mumby and Putnam 1992: 466). In the political sphere, the supposed emotionality of women has been used to explain their exclusion from positions of power as well as legitimize their disadvantaged social and economic position. To take just one example, women's de facto exclusion from the office of the American Presidency has long been legitimized in part via an appeal to the notion that women are "too emotional" (Lutz 1986: 299) and thus less capable of objectivity and sound judgment. The assumption is that being emotional is tantamount to being overcome with feeling, and therefore incapable of logical thinking or sensible behavior. In part, this is because emotion is thought to entail bias and thereby render subjects blind to rational judgments that they ought to make.

As noted in reference to Rousseau's account, this dualism of emotion and rationality bears a strong connection to traditional gender roles, assumptions about the separation between private and public spheres, and the notion that men and women have distinct roles to play in society. The family tends to be understood primarily as an emotional unit, headed up by women, while the workplace is governed by the calculated means-end reasoning of men. This is related to the traditional view that women's bodily nature, together with their "natural" tendency to be more expressive of their emotions, makes them well-suited to child rearing and other caretaking roles. But in addition, as noted above, it also implies that women are less adept when it comes to engaging in various modes of cognition, including the sort of decision-making, reflective judgment, and abstract thought that is required in the public sphere. This is because the same emotionality that makes them well-suited to empathic labor within the home also renders them prone to "react oversensitively to the rough and tumble of commerce and workforce discipline" (Lutz 2001). Thus, the differential assignment of reason and emotion functions "to bolster the epistemic authority of the currently dominant group" (Jaggar 1989: 158) and to reinforce gendered divisions of labor.

Some may find the ideas presented thus far to be an oversimplification of the range of views that Philosophers and other theorists have presented about rationality and emotion and their relation to gender difference. After all, the split between reason and emotion was not absolute even for the Greeks. On Plato's account, the emotions provide indispensable motive power that needs to be channeled appropriately with the help of reason. Numerous theorists have noted that without the "unruly horses" of desire and emotion there to *move* the chariot along, the skill of the charioteer would be worthless and the chariot would not go anywhere. Still, Plato's conception of the tripartite soul brings with it the notion that emotion and reason are distinct and separable faculties, and that emotion always should be controlled by reason. Moreover, these ideas cannot be dismissed merely as outdated modes of thinking. In contemporary political discourse, the concept of emotion still continues to be employed as a category to talk about that which deviates from dominant understandings of what is sensible. In this society, women, people of color, children, and people from lower socioeconomic classes have been labeled emotional, with the implication that they are less than fully rational or sensible (Lutz 1986: 292). Overcome by emotion, they are unable to pursue their goals effectively, and this both explains their exclusion from positions of power as well as legitimizes their disadvantaged social and economic position. In addition, as I will discuss in the next section, the notion that emotion and reason are separable and distinct continues to shape contemporary thought and discourse about the nature of cognitive functioning. Although some philosophers have begun to pay more positive attention to emotion, the structure of the discussion still "presupposes that theoretical, detached cognition epitomizes the nature of our most basic relationship with the world" (Ratcliffe 2002: 294–295).

In response, some feminists may want to insist that that the sexes are in equal possession of reason and that women should be admitted to full participation in the public sphere of work and politics. But while affirming women's rationality probably is quite important, merely insisting that women are rational *just like men* "seems implicitly to accept the downgrading of the excluded character traits traditionally associated with femininity" (Lloyd 1984: 104), i.e. emotion and bodily existence. Another possible response is to affirm the value and strengths of distinctively feminine traits (Lloyd 1984: 105). Some recent works in pop neuroscience, for example, have explored the possibility that women's brains give them particular talents and abilities, ones which differ from those of men (Brizendine 2006; Schulz 2006). Similarly, some feminist theorists have suggested that emotions serve a function that never could be achieved by cooler, abstract powers of reason. Such theorists accept that women are more emotional, but celebrate their "closer approximation to the natural state of human beings, which is to be in relationship with others" (Lutz 1986: 301). In the feminist literature on moral psychology, for example, care ethicists drawing on the work on Carol Gilligan (1982) often have accepted that men and women tend to reason differently about moral matters. However, such theorists deny that the male tendency to focus on abstract rules and principles of justice is superior, insisting that women's tendency to put more emphasis on care for others often is a more reliable guide to moral decision-making. Jaggar (1989), likewise, famously characterizes emotionality as a source of women's strength. This is because women and members of other subordinate groups are more likely to experience "outlaw emotions," which constitute an "epistemic resource" and are crucial for developing a critical perspective on the world. Such emotions come into play, for example, when someone feels angry when she hears a sexist or racist joke. The idea that emotions can serve as an epistemic resource is, no doubt, a very important insight. However, as Jaggar no doubt readily would acknowledge, it seems clear that men, too, can experience outlaw emotions in response to sexist jokes, and it would be a mistake to accept the assumption that women are naturally more emotional than men. It is not enough to say that "emotions constitute a way of knowing that differs from but complements traditional rationality" (Mumby and Putnam 1992: 480). The danger is that any implicit acceptance of women's emotionality will allow male norms to persist, and simply will perpetuate the idea that the female is complementary to, *but different* from, the male. The assumption that women and men are fundamentally different will be preserved, and the association of rationality with maleness will remain unchallenged despite these theorists' noble intentions.

Just as I reject the notion that women naturally are more emotional and that men naturally are more rational, I think it would be a mistake to deny the important contribution made by emotions. However, the central problem with both of these responses is that they accept the emotion/reason divide to some extent. And this adherence to the notion that reason and emotion are distinct faculties inadvertently lends support to the assumption there are clear and fundamental differences between men and women. Perhaps it is not surprising that this picture of distinct faculties gets incorporated even into feminist theorizing given that the "dichotomous categories of 'cognition' and 'affect' are themselves Euroamerican cultural constructions, master symbols that participate in the fundamental organization of our ways of looking at ourselves and others both in and outside of social science" (Lutz 2001). Because the use of these dichotomous categories is so widespread, it becomes difficult to challenge them or even to speak about how they might be challenged.

It is unsurprising, then, that what we see in much of the contemporary work being done in Philosophy of Psychology involves an implicit acceptance of the reason/emotion divide. Now, it is true that in their accounts of cognitive functioning, many theorists do recognize that emotions can play an important role and thus have moved past viewing reason simply as superior to emotion. However, as I will show in the next section, even these theorists often take something more akin to a "separate but equal" approach that is somewhat reminiscent of Rousseau's notion that reason and emotion are "complementary." The reality, I will suggest, is that emotion and reason/cognition are neither opposed to nor fundamentally different from each other, nor are they clearly separable. Acknowledging this is central to gaining a better understanding of how various forms of cognition actually work. But in addition, such acknowledgement is crucial for feminist theorizing given that the ideologies surrounding gender and the supposed reason/emotion binary reinforce each other and have serious implications for widespread cultural assumptions about where weakness and inferiority are to be found (Lutz 1986: 300). Perhaps, then, acknowledging that there is no sharp divide between emotion and reason may help to demonstrate that there is no sharp divide between female and male modes of cognitive functioning. It is untrue that women are more emotional, just as it is untrue that emotions are a source of weakness, and this is because in instances of effective cognitive functioning among all ordinary adult human beings, reason and emotion are inseparable and mutually interdependent.

7.3 The Reason/Emotion Divide in Contemporary Accounts of Cognition and Emotion

According to the intellectualist tradition commonly associated with Plato and Descartes, thought can be treated as a kind of computation. Computer programs are algorithms, and if we are computers, then it must be possible to uncover what sort of program we are running to perform various cognitive tasks. Cognition is a matter of processing information and involves distinct modules, each of which performs a particular function.

Along these lines, the theory-theory (TT) of social cognition asserts that ordinary human adults have an innate mind-reading module that allows them to make of use of causal-explanatory generalizations (a tacit folk psychological theory) to "read minds" and thereby make sense of and predict other people's behavior. Relying on a set of causal laws that interrelate inputs, internal states, and behavioral outputs, they are able to attribute mental states and make inferences about an agent's future behavior (Goldman 1995: 186). The capacity for employing this folk psychological theory depends on higher-order cognitive processes such as theorizing and inference-making and ordinarily surfaces at about four years of age. Simon Baron-Cohen (1999, 2002) makes sense of empathy in this way, as centering upon a capacity to mentalize, identify another person's emotions and thoughts, and predict his or her behavior. In his view, we theorize "an enormous amount, as a natural way of thinking about why people do what they do" (Baron-Cohen 1999: 177). Empathizing involves "an imaginative leap in the dark, in the absence of much data" and renders a causal explanation that "is at best a 'maybe'" (Baron-Cohen 2002: 248) For example, the thought that "Maybe she didn't phone me because she was feeling hurt by my comment" illustrates the ordinary human ability to take these sorts of imaginative leaps, which Baron-Cohen claims is our "most powerful way of understanding and predicting the social world" (Baron-Cohen 2002: 248). It is noteworthy that neither emotion nor a bodily mode of relatedness plays a central role in this account of empathy. Instead, Baron-Cohen's account treats social cognition as a wholly heady, intellectual process that often takes place at a distance and without the help of affectivity or bodily feeling.

The "simulation-theory" (ST), on the other hand, suggests that mind-reading is a matter of mentally simulating another person's mental life. The ability to attribute mental states depends on our having an ability to simulate others' mental processes, which itself cannot be explained in terms of the possession of a body of theoretical knowledge. According to some proponents, it is reasonable to suppose that when someone is trying to work out another person's motives and intentions, they can make a good start on doing so by considering what their intentions and motives would be if they were situated as they take this other person to be situated. Our own mind thereby serves as sort of model that we can use simulate the other person's mental state and then predict what he or she will do. Some theorists have described it as a matter of feeding another person's mental states into one's own practical reasoning mechanism and then ascribing the output to this other person. According to Goldman (1995), one first imagines being "in the shoes" of the agent, and pretends to have the same initial desires or beliefs that the available background information suggests the agent has. Next, one feeds these pretend states into some "inferential mechanism" and allows this mechanism to generate outputs. These output states can be viewed as "pretend" or surrogate states that allow one to predict what the other person will do. During this simulation procedure, there is a sense in which one allows one's own psychological mechanism to serve as a model for the other person's mind (Goldman 1995: 189).

The two theories of mind are similar insofar as they both depict mind-reading as a matter of inferring people's mental states on the basis of their outward behavior. However, while TT maintains that mind-reading is primarily a matter of cognition and intellect, ST leaves more room for the notion that we draw heavily from our emotions, desires, and practical skills. Insofar as it recognizes the role that emotions such as empathy play in our attempts to understand others' mental states and behavior, ST is better positioned than TT to accommodate the possibility that emotion plays a central role in social cognition. However, both of these cognitivist approaches mistakenly involve a spectatorial and detached view of interpretation and interpersonal understanding that ignores the crucial role of second-person interaction and bodily engagement. Both treat social cognition as a heady, intellectual process that relies on bodily feelings and emotions only instrumentally (if it relies on them at all). And both assume that our ordinary understanding of mindedness centrally consists of an ability to attribute internal mental states in order to explain and predict others' behavior, rather than a spontaneous and more intuitive mode of second-person engagement. Causally efficacious inner states are construed as tucked away inside people's heads, in the realm of the mind. Because we cannot directly perceive others' thoughts, feelings, or intentions, we need some special cognitive process (whether theorizing or simulation) that will allow us to infer the presence of such mental states. Any discussion of how external signs such as bodily comportment, expression, gesture, and movement might play a direct and integral role in helping us to understand others seems to be absent. In short, these theories describe mind-reading, inference, and simulation as if they were thoroughly intellectual, brainy, but otherwise non-corporeal processes. Insofar as they say little about the role of embodied interaction, bodily attunement, immediate emotional resonance, or direct perception of others, such accounts seem to be in line with the longstanding assumption that cognition and intellect are distinct from emotion and bodily feeling, and that it is the intellect *rather than* emotion and bodily feeling that is central to thought and understanding.

The presumed separation between emotion and reason also surfaces in many prominent accounts of decision-making and moral judgment. Theorists sometimes have described the process of making choices, aligning choices with goals, and taking steps to achieve those goals as if it were divorced from feelings, sensations, and emotions. For example, according to Simon, individuals within organizations act "rationally" by relying on means-end reasoning to make a series of incremental decisions, and emotions produce "bad long-run consequences for the organization" (Simon 1989: 35). Perhaps this could be dismissed as an extreme, overly simplified view that is not widely held. However, the reason/emotion divide is presumed even in more nuanced accounts of decision-making. For example, Greene and colleagues (2004: 389) distinguish between two kinds of decision-making process, which they characterize in terms of two distinct and "mutually competitive" psychological subsystems. Interestingly, they do acknowledge that this distinction between emotion/affect and cognition/reasoning may be somewhat artificial. However, it is clear that it plays a central role in their account. They maintain that while System 1 generates judgments via evolutionary ancient, domain-specific social emotional-dispositions, System 2 involves control processes that can serve to override the prepotent emotional dispositions of the first system. According to Greene and colleagues, System 2 processes are more effortful and are associated with the recognition of a conflict between competing considerations. In cases where an intuitive answer conflicts with a principle based answer, subjects show increased activity in brain regions associated with the recognition of cognitive conflict, and decreased activity in regions associated with emotional processing. Moral judgments are driven by either one system or the other, and reflective processes are crucial for overriding intuitive responses. Greene and colleagues thereby characterize these two types of process as being fundamentally in *competition* with one another, and make it clear that System 2 should be in control. Similarly, Haidt (2007) distinguishes between "moral reasoning" and "moral intuition" and characterizes moral reasoning as a controlled, cool, and relatively non-affective process that consists of processing information about people and their actions in order to make a judgment or decision. Moral intuition, on the other hand, is a "fast, automatic, and (usually) affect-laden [process] in which an evaluative feeling of good-bad or like-dislike (about the actions or character of a person) appears in consciousness without any awareness of having gone through the steps of search, weighting evidence, or inferring a conclusion" (Haidt 2007: 998). Such theories reflect a persisting tendency to suppose that conscious reasoning can be marked off clearly from the operation of intuitive-emotional responses, that these two sorts of processes may be in tension, and that it is better for reasoning to be in control.

Even theorists who acknowledge that affective, intuitive processes somehow are integrated with more reflective processes typically view the two as separable and distinct. Indeed, "the partitioning of cognitive processes into intuitive and deliberative kinds is a theme that has become increasingly popular in the cognitive sciences over the past three decades" (Craigie 2011: 57). The "dual-process" account of Kahneman and Frederick (2002), for example, characterizes decision-making as involving both "System 1" and "System 2" processes. While System 1 processes occur rapidly and automatically, without effort, and are inaccessible via introspection, System 2 processes are deliberately controlled, slow, and transparent, and involve conceptual content that can serve to decontextualize judgments. The main advantage conferred by System 1 processes is that they can produce responses quickly and efficiently without requiring much in the way of cognitive resources. However, because such processes are not flexible enough to take into account factors that can distort impressions, the reflective processes of System 2 play a crucial role in monitoring, endorsing, modifying, or overriding System 1 responses. Because each system has its own particular strengths and weaknesses, competent decision-making requires the integrated functioning of the two. Kahneman and Frederick's model thus allows for both intuitive and more deliberative processes to play a role in decision-making. One might say that these two sorts of processes can be viewed as *complementary*, which resonates with Rousseau's depiction of the emotional minds of women as complementary to the rational minds of men.

Drawing on Kahneman and Frederick's model, Craigie (2011) argues that both intuitive/emotional processes as well as more effortful, reflective ones play a significant causal role in moral cognition. She maintains that one key factor that Greene and colleague's model leaves out is the extent to which moral intuitions (System 1 processes) are amenable to modification by reflective (System 2) processes (Greene et al. 2004: 60). As proficiency and skill are acquired, what was once a System 2 process becomes more automatic and habitual. Intuitive responses come to reflect, at least to some degree, the output of more reflective, deliberative processes. For example, over time, deliberate control of racial prejudice can become an intuitive, automatic response (Craigie 2011: 60). While this sounds accurate, such an account remains very much consistent with the idea that reason should "hold the reins" and steer emotion in the proper direction. What is absent from this proposal is the recognition that reflective processes can be modified by emotions and that rather than detracting from moral cognition, this emotional influence may very well be crucial for effective cognitive functioning. Moreover, by treating intuitive/emotional processes and more reflective, deliberative processes as two distinct systems, Craigie's account preserves the notion that these are two fundamentally different and separable modes of processing information.

In his more recent work, Greene (2007) has refined his claim that cognitive processes are in competition with emotional ones. He goes so far as to say that he does not think that there is a sharp distinction between cognition and emotion, and that he is sympathetic with Hume's claim that all moral judgment "must have" an emotional component (Greene 2007: 41). However, he maintains that different kinds of emotion are involved in different kinds of moral judgment. On the one hand, the emotions that are essential to deontological moral judgments are alarm-like: in effect, they issue commands by virtue of being vivid and motivationally demanding. The emotions essential to consequentialist moral judgment, on the other hand, are currency-like: in effect, they say, "Such and such matters this much. Factor it in" (Greene 2007: 64). Here Greene seems prepared to acknowledge that cognitive and emotional processes are integrated rather than being in competition with each other. Nonetheless, he makes it clear that he is still committed to the cognition/emotion dichotomy. He maintains that consequentialist judgment is inescapably cognitive, and that it cannot be implemented in an intuitive, emotional way (Greene 2007: 65). This is because "cognitive" representations are inherently neutral, while "emotional" representations are behaviorally valenced representations that trigger particular behavioral responses or dispositions (Greene 2007: 40). One can represent CAR, for example, in a behaviorally neutral or "cognitive" way. Likewise, Greene maintains that what occurs in the context of impersonal, Utilitarian moral judgment is a "cognitive," cost-benefit analysis that does not depend on emotions in the way that deontological judgments do. Thus, Greene's more recent account still treats the "emotional" and "cognitive" as fundamentally different, and as resulting in distinct types of moral judgment. It draws a clear contrast between moral cognition that is implemented in an "emotional" way and moral cognition which is implemented in a "cognitive way."

Interestingly, the same sort of binary thinking plays out even in characterizations of emotion itself, which surfaces in the philosophical debate between so-called cognitive and bodily theories of emotion. Generally speaking, cognitive theorists characterize emotions as nothing but certain kinds of evaluative judgments or appraisals; while feeling theorists define emotion in terms of bodily feelings or an awareness of certain physiological changes. The early Solomon (1976) denies that "feelings" are a central feature of the emotions, arguing that feeling is the ornamentation of emotion rather than its essence. According to his cognitive account, fear cannot be a mere bodily feeling because the physiological contour of fear is no different from the physiological contour of anger. Moreover, while fear can be appropriate or inappropriate, feelings cannot, and it would be a mistake to identify emotions with objectless feelings or view them as analogous to headaches. Instead, emotion should be understood as an evaluative judgment about one's situation, a "personal evaluation of the significance of [a particular] incident" that projects our values and ideals (Solomon 1976: 126). A central motivation for the feeling theory, on the other hand, is the commonsense idea that bodily feeling or affectivity is essential to emotion, which seems to flow directly from the phenomenology of emotional experience. Along these lines, Whiting (2009) identifies emotions with feelings and characterizes emotions as non-cognitive. He maintains that fear, for example, should be identified with the "awful edgy sensation that assails my guts and limbs" (Whiting 2009: 287).

This debate stems, in part, from a long-standing assumption that cognition and thought are abstract, intellectual, disembodied processes, and that bodily feelings are non-intentional and have no representational content. Indeed, one interesting thing to note about the debate between cognitive and feeling theories is that despite their divergent views on what defines emotions, both sides seem to agree that bodily feeling is non-cognitive and non-intentional. To see this, note that like Solomon, Whiting denies that bodily feelings have objects or a representational structure. He maintains that the edgy feeling that pervades my guts and limbs upon seeing a tiger just *is* the emotion (fear), and that the representation of a tiger is not part of this experience. Likewise, the representation of some sort of offense is no part of what it is like to undergo the irritable feeing that defines the emotion of anger. In order to account for the fact that "fear-of-tiger" and other object-directed emotions do seem to involve mental representations, Whiting recommends that such mental states be viewed as composite states consisting of an emotion (a feeling) *and* a thought (with representational content). Fear-of a tiger, then, is a composite mental state that consists of an emotion (the feeling of fear) *plus* a thought about a tiger. This compound mental episode comprises a feeling together with a representation, but the sensation "presents only as a non-representational feeling state" (Whiting 2012: 100). To make sense of how the emotions and thoughts that together constitute object-directed mental states relate to each other, Whiting suggests that "the emotion follows on immediately from the thought and that emotion is directly caused by that thought" (Whiting 2012: 288). Emotions do not have an intentional nature or structure; in his view, it is thoughts that represent, not feelings. Although they disagree about what constitutes the essence of emotion, both Solomon and Whiting appear to be in agreement that thought and feeling are detachable components. Insofar as bodily feelings lack intentional structure and representational content, they are clearly distinct from thought and appraisal.

Even hybrid theories that aim to bring together the affective/bodily and appraisal dimensions of emotion tend to view the two as separable. Prinz (2004), for example, follows in the tradition of William James' well-known feeling theory, which depicts emotions as perceptions of physiological responses (i.e. changes in the autonomic, hormonal, musculoskeletal, and motor systems). He denies that emotions require conceptualized appraisals or necessarily are constituted by propositional attitudes. In his view, a state is cognitive just in case it involves representations that are controlled by structures in executive systems, and which are activated, maintained, and manipulated by the organism rather than by the environment. Because he believes that emotions are passive and often not under the organism's control, he concludes that most of the time, emotions are not cognitive. Thus, in Prinz's view, emotions are not cognitive appraisals, but rather mental states that detect and register bodily changes, represent objects or events as having some bearing on one's interests and concerns, and thereby track organism-environment relations. For example, fear registers an array of bodily changes, tracks danger, and represents events as posing a threat to one's interests and concerns. Danger is a relational property, for "something can be dangerous only to some creature or other, and whether or not something is dangerous depends on the creature in question"

(Prinz 2004: 63). Drawing from the work of Paul Lazarus, Prinz describes these relational properties that pertain to well-being as "core relational themes," and maintains that emotions track these core relational themes by registering changes in the body (Prinz 2004: 68). Insofar as certain bodily changes reliably co-occur with certain organism-environment relations (core relational themes), emotions use our bodies to tell us how we are faring in the world.

In summary, Prinz maintains that the bodily dimension of emotion is passive, that the body is extrinsic to the process of appraisal, and that most of the time, emotions are not cognitive. His account seems to carve an emotional episode into input, bodily reaction, and subsequent cognition. His embodied appraisal theory says that bodily changes come first and then a separate appraisal process comes along subsequently to monitor these bodily changes. Thus, on his view, the bodily changes and the appraisal process are separate components of emotion and it is the appraisal component which does all the "smart" interpretive work (Colombetti 2014: 91). The upshot is that like so many other emotion theorists, Prinz construes cognition as disembodied and treats bodily affectivity as detached from the cognitive component of emotion. His account thereby perpetuates the alleged division between affectivity and cognition despite its potential to bring together the cognitive and bodily elements of emotional experience.

7.4 Moving Beyond the Reason/Emotion Dichotomy

In the previous section, I attempted to show how the presumption of a reason/emotion divide continues to shape contemporary discourse in the field of philosophy of psychology. However, there are some notable exceptions to this trend. In fact, a growing number of thinkers (Damasio 1994; Evans 2002) have begun to acknowledge the importance of the emotions, and some maintain that emotion is integral to practical reasoning and decision-making, especially when these processes involve complex personal and social matters. This is because emotions involve various patterns of discrimination, focus, and salience that allow us to see what sort of action is fitting or appropriate given what we care about (Zemach 2001). It is because our emotions are particularly good indicators of import that so many of our difficult decisions seem to be grounded ultimately in an appeal to "gut instinct." Along these lines, Matthew Ratcliffe (2002, 2008) has argued that our cognitive engagements with the world are essentially bodily and felt and that even philosophical outlooks may owe a great deal to one's bodily and affective disposition or temperament.

Many of the philosophers I discuss below work in the field of *embodied and enactive cognition*, which emphasizes ways in which the body and emotions make an active contribution to our understanding of our surroundings. These theorists have presented accounts of social understanding, moral judgment, and emotion that emphasize how the bodily and affective aspects of these processes are thoroughly bound up with their intellectual and cognitive aspects. To my knowledge, these theorists (myself included) rarely explicitly acknowledge the feminist themes in their work. However, I maintain that by challenging the dichotomies of reason/emotion and intellect/body, they are making an important contribution (whether intentionally or unwittingly) to the dismantling of binary and essentialist ways of thinking about gender. As noted in the introduction, this is because by challenging these dichotomous categories, these theorists help to undermine the notion that men and women have fundamentally different modes of engaging with the world. Whether or not they realize it, these theorists have made an important contribution to feminist philosophy of psychology.

For example, the accounts of social cognition presented by theorists such as Shaun Gallagher, Matthew Ratcliffe, Hanne De Jaegher, and Ezequiel Di Paolo all highlight the ways in which emotion and bodily feeling are thoroughly bound up with cognitive processing. Gallagher's proposed "interaction theory" emphasizes the importance of embodied interaction and direct perception. He points out that in ordinary social interactions, I typically do not occupy an observer position, but rather come to understand others via perception-action loops and through the various things I am doing with or in response to others (Gallagher 2008: 168). Communication between individuals often occurs in an embodied manner, via their postures, gestures, and facial expressions. In his view, "before we are in a position to theorize, simulate, explain or predict mental states in others, we are already in a position to interact with and to understand others in terms of their gestures, intentions, and emotions, and in terms of what they see, what they do or pretend to do with objects, and how they act toward ourselves and others" (Gallagher 2005: 91). We do not see facial contortions and then infer that someone is feeling sad, but rather have a direct perception of his sorrow, insofar as it is "written all over his face." In short, rather than having to infer mental states, "we perceive them in the movement and expression of the other's body" (Gallagher 2005: 90). Likewise, according to Ratcliffe (2007), we typically are able to perceive the goal structure of actions in the gestures, expressions, and movements of others. Engaging with another *as a person* involves adopting a personal stance, comprised of affective and bodily relatedness, and it is these social encounters where we address another as "you" that provide the greatest insight into interpersonal understanding (Ratcliffe 2007: 23). Both theorists suggest that the ability to understand others rests on a form of embodied practice that is emotional, sensory-motor, and perceptual.

It is worth noting that according to Gallagher's proposed account, the interaction between infant and caregiver serves as the basis for all social cognition. The first set of capacities upon which more sophisticated modes of social understanding are built are pre-theoretical and nonconceptual, embodied in intersubjective perception, and manifest from early infancy. Following Trevarthen (1979), Gallagher refers to this set of basic capacities "primary intersubjectivity." The existence of such capacities is supported by the fact that young children appear to be highly sensitive to the feelings, attitudes, and interests of others long before they are able to pass the false-belief test. Prior to age three, children already have a sense of what it means to be an experiencing subject and can identify other entities in their environment that are subjects similar to, and yet different from, them. An infant is able not only to pick out a human face from the crowd of objects in its environment, but also to imitate the gesture it sees on that face and use facial gesture to provoke responses from others. In neonate imitation, the newborn demonstrates proprioceptive awareness of her own body, the recognition of a distinction between self and non-self, and also the recognition that the other is in fact the same sort of entity as herself. Infants in their first hours of life are drawn to faces and quickly become quite skilled at detecting what others are feeling. They also are attracted to their mother's voice, and have what might be called an "eye direction detector" and an "intentionality detector." Infants as young as 6 months perceive grasping as goal-directed; at 9 months they follow the other person's eyes and start to perceive various body movements as goal-directed movements; and at 10–11 months they are able to parse some kinds of continuous action according to intentional boundaries (Gallagher 2008: 166). In short, infants look to others' bodies and expressive movements in order to discern their intentions and emotions and make sense of their behavior.

This ability to interpret bodily movement as goal-directed clearly does not require the capacity for theorizing, since it surfaces at such a young age, and appears to operate quickly and automatically. Such evidence supports the idea that infants are capable of a perceptually-based, embodied understanding of the intentions and dispositions of other persons. Gallagher maintains that this sort of embodied understanding serves as the basis for all face-to-face intersubjective experiences, and even comes into play in those instances when we explain or predict what other people believe, desire, or intend in more intellectually-sophisticated ways. By treating the interaction between infant and caregiver as the basis for all social cognition, his account challenges the notion that understanding others is a purely intellectual, unemotional, disembodied achievement. Interestingly, it also locates the beginnings of social cognition in a sphere traditionally assigned to women: the sphere of caretaker-child relations.

The account presented by De Jaegher and Di Paulo (2007) likewise emphasizes how social cognition emerges and is modulated by specific patterns of bodily engagement and interaction among the parties involved. This is because social cognition involves ongoing engagement and *coordination*, which is the non-accidental correlation between the behaviors of two or more coupled systems. This involves coherence or matching of behavior over and above what is expected given what those systems are capable of doing. Through the coordination of intentional activity, the way that each individual understands a situation is mediated by the sense-making activities of the other person(s) involved in the encounter (De Jaegher and Di Paolo 2007). These are instances in which an action is directly and spontaneously apprehended, and one immediately responds with a similar or complementary gesture that allows the other person to see that one has recognized the meaning of her action (Ratcliffe 2007: 141).

Examples of coordination in the realm of human activity include synchronization, mirroring, anticipation, and imitation, all of which are displayed by infants from a very early age. Indeed, second-person interactions between infants and caregivers are centrally characterized by reciprocation of affect and emotions, and infants naturally become distressed when others stop interacting with them (Striano and Reid 2006: 471). The infant who reduces her smiling and gazing, and then attempts to reengage her social partner through smiles and vocalization, is in effect inviting her social partner to communicate with her and coordinate his behavior. Another striking example is how when see a smiling face or some other facial gesture, we immediately and involuntarily attune to it with an enactive, mimetic response (Gallagher 2008). And perhaps even more striking is the phenomenon of infectious laughter, since it is more obviously emotionally driven and fully embodied. When other people laugh, we are inclined to laugh too; when other people are anxious, we become anxious; and when others are at ease, this puts us at ease. These are just a few very common examples of emotional contagion or what some theorists have described as "automatic emotional resonance" (Hatfield et al. 1994: 2). In addition, interaction partners mirror each other's movements, anticipate them, temporally synchronize or desynchronize them, and alter them in accordance with the intricate to-and-fro glances, utterances, and gestures that occur between them (De Jaeagher 2009: 539). Such perceptual-motor coupling is particularly evident in situations of intricate bodily coordination, such as dance or sport, in cases where there is a great deal of bodily attunement and mutual modulation. In such instances, motor mimicry, matched or coordinated body positions, and complementary movements or gestures helps the interactors to understand one another and also to make sense of their surroundings.

All of these accounts depict social cognition as fully embodied and emotionally driven, and thus move us away from the notion that social understanding centers upon disembodied, detached intellectual processes. Such accounts open up space to say that empathy typically is a matter of coming into affective bodily contact with other persons' mental states, so that one literally *feels with them*. Social understanding is indeed a cognitive process, but one thoroughly intertwined with emotion, feeling, and bodily engagement. Understanding other people's mental states and behavior relies centrally and necessarily on the embodied, emotional second-person interaction process itself.

Likewise, in my own work on moral cognition (Maiese 2013), I have argued that affective and reflective processes posited by dual process theorists are not only integrated, but also inseparable and *mutually interdependent*, and that only together do they allow for full-fledged decision-making and moral judgment. What I call "affective framing" is a spontaneous, non-inferential, and pre-reflective way of discriminating, filtering, and selecting information that involves bodily attunement and feelings of subjective import. Drawing from the work of Husserl, Thompson (2007) describes affect as the *allure* of consciousness, the pull that an object given to consciousness exercises on the ego. Allure motivates attention, and implies a "dynamic gestalt or figure-ground structure" whereby some objects emerge into affective prominence, while other objects become unnoticeable (Thompson 2007: 374). This sort of allure focuses one's attention and "conditions and circumscribes the kind of cognitive engagement one is able to have with the world" (Ridley 1997: 174). The way in which an agent interprets an object, situation, or event, and also highlights some considerations while overlooking others, carves out the "starting points" for deliberation, thought, and evaluative assessment. The notion of affective framing is meant to capture the idea that *all* intentionality is affective intentionality, and that the mind's directedness toward the world arguably is *never* a cold, detached, purely intellectual affair. This is because bodily feelings always are implicated in our awareness of things outside the body, so that it is only through the felt body that one can affectively apprehend something as significant. Thus, emotions can be understood as constituting interpretive frameworks, patterns of attention, determinate patterns of salience, and inferential strategies (Jones 2004: 335–336). Affect and bodily feeling reveal saliencies that we might not otherwise recognize with the same speed and reliability (Goldie 2004) and thereby attune us to the world. Just as observation directs and partially defines emotions, emotional attitudes influence what features of the world are selected as important (Jaggar 1989: 160). Affective framing allows for the very fine-grained attentional focusing that is needed to highlight specific details in accordance with what we care about, what moves us, and what we *feel* is important. This is crucial for ordinary decision-making and moral evaluation, which are themselves extremely fine-grained processes that require the subject to attend to some particular features of the case at hand rather than others.

Thus, there is good reason to think that our capacity for moral cognition *depends* on, and in fact crucially involves this capacity for immediately focusing our attention on what we care about. Even observers who do not recognize that they are influenced by their emotions approach everyday decision-making and moral assessment with a particular worldview and patterns of attention and response that are infused with affect and unavoidably subjective. However, rather than detracting or distorting the reasoning process, these feelings are inseparable from effective cognition and play a central role in allowing processes of decision-making and moral cognition to take place. In short, what we care about focuses our attention and serves as an extremely efficient and effective way of delineating which factors are to be included in decision-making or moral evaluation. By way of affective framing, we are able to gauge salience and significance and then direct our attention to those features that our bodily feelings have underscored as weighty so that we may consider them further. If this is true, then the usual distinction that dual-process theorists make between deliberation and conscious reasoning on the one hand, and bodily affect and emotion on the other hand, is mistaken. A central part of reasoning processes is an immediate, affective, pre-reflective, non-inferential mode of interpretation that involves affectivity and bodily feelings. This tight link between cognition and emotion is supported by the fact that among infants, "attention reactions, (the immediate focusing of attention on newly appearing stimuli) involve emotions such as interest, fear, and surprise" (Ciompi 2003: 188).

Insofar as emotions focus attention and highlight particular facts as ones that matter and should be taken into account, they shape our overall understanding of our surroundings, and thus crucially inform both our intuitive responses as well as our more reflective thought processes. Therefore, contra Greene's (2007) account, both deontological and consequentialist moral judgments should be understood as simultaneously cognitive and emotional. Even representations such as "five lives saved are better than no lives saved," which seem to be relatively emotionally cool, may very well be intrinsically motivating, and thus behaviorally valenced (Orsi 2012: 12). Granted, the utilization of this sort of representation may not be emotionally intense. Still, emotion arguably is part of what underlies the characteristically Utilitarian realization that by harming one we can save others, and that this counts in favor of harming one. The fact that subjects care for the well-being of those who might be saved has an impact on their patterns of attention and their capacity for gauging salience, and thereby plays a central role in their judgments. In the effective moral cognitive functioning of ordinary adults, reflective reasoning processes and affective processes are thoroughly intertwined and interdependent, and emotion is an integral part of the reasoning process.

Lastly, a growing number of emotion theorists have begun to challenge the alleged split between the bodily/affective and cognitive components of emotion. For example, Furtak (2010: 55) maintains that much headway could be made if theorists were to stop assuming that the cognitive and bodily aspects of emotions can be separated, or that one aspect has some kind of logical or causal priority over the other. Likewise, Colombetti and Thompson (2008) have argued that emotions are simultaneously bodily and cognitive-evaluative and that bodily feelings are constitutive of the sense of personal significance that emotions involve (58). Thus, there is good reason to think that the appraisal and bodily aspects of emotion are *constitutively interdependent* (Colombetti 2007, 2011). Rather than being extrinsic to or a mere byproduct of appraisal, bodily arousal is part of the very experience of appraisal (Colombetti 2011). For example, an interview-situation is evaluated and experienced as anxiety-provoking through the state of one's body. Likewise, when one evaluates something as a loss, the appraisal already is imbued with feelings of sadness. Evaluating the world and responding to it emotionally are not distinct processes, and the process of appraisal overlaps with the bodily components of emotion (Colombetti 2014: 112). This is because appraisal involves recognizing something as relevant or salient in relation to one's own goals, capacities, and needs, and because the body itself often serves as a vehicle of salience. A subject's physiological state and bodily feelings constrain the range of cognitions available to her and influence how she interprets situations, facts, and states of affairs. Central to this account is the notion that emotions are *enactive*, i.e. that they are a matter of active appraisal. A subject in the midst of emotional experience does not simply passively receive information from her environment, but instead actively participates in the generation of meaning, and she does this in large part by becoming selectively attuned to her surroundings in and through her affectively aroused body.

This understanding of emotion is supported by the empirical fact that the brain and bodily systems responsible for appraisal largely overlap with the systems for arousal (Lewis 2005). The sub-personal processes that underlie appraisal and emotion are a distributed network of self-organizing and mutually influencing brain and bodily processes. Together with the amygdala, bodily arousal and endocrine activity help to maintain an organism's homeostatic equilibrium, enhance attention, and prepare the individual for action. Thus, the process of appraisal is best characterized as a broadly organismic activity, one which overlaps with the bodily component of emotion (Colombetti 2014: 112).

By emphasizing the interdependency between bodily and cognitive aspects of emotion, this sort of account undermines the notion that bodily feelings lack any sort of representation content; it also challenges the assumption that appraisal is a "heady" intellectual process that takes place apart from bodily arousal. In addition, this enactive account of emotion moves us away from the notion that emotions are passive, and explores how the bodily elements of emotional experience might make an active and significant contribution to the appraisal process.

7.5 Conclusion

Acknowledging that emotion and the body are constitutive of the capacity to understand and act adaptively is crucial for our understanding of various modes of cognition (Colombetti 2011). But in addition, challenging the presumed distinctions between reason/emotion and intellect/body is crucial for feminist theorizing. I have argued that this is because challenging the dichotomy between rationality and emotionality serves to debunk the notion that certain ways of engaging with the world are either characteristically masculine or characteristically feminine. If it is true that emotion and affectivity are always and everywhere thoroughly bound up with cognitive processing, then it makes little sense to speak of emotion and reason as if they were distinct and separate faculties. Instead, emotion/affectivity and reason/cognition are *constitutively interdependent* and not at all clearly separable, and all healthy modes of cognition involve what we would call an emotional element. It makes little sense, then, to suppose that men and women understand and experience their surroundings in fundamentally different ways. There is no "man of reason" and there is no "woman of emotion." Instead, the effective cognitive functioning of all ordinary human adults is fully embodied and emotionally driven. No doubt further work needs to be done in philosophy of psychology to develop this sort of account of cognition. I maintain that this is precisely the sort of work that a feminist philosopher of psychology ought to be doing.

References

Baron-Cohen, S. (1999). Can studies of autism teach us about consciousness of the physical and the mental? *Philosophical Explorations, 3*, 175–188.

Baron-Cohen, S. (2002). The extreme male brain theory of autism. *Trends in Cognitive Science, 6* (6), 248–254.

Brizendine, L. (2006). *The female brain*. New York: Broadway Books.

Ciompi, L. (2003). Reflections on the role of emotions in consciousness and subjectivity, from the perspective of affect-logic. *Consciousness and Emotion, 4*(2), 181–196.

Craigie, J. (2011). Thinking and feeling: Moral deliberation in a dual-process framework. *Philosophical Psychology, 24*, 53–71.

Colombetti, G. (2007). Enactive appraisal. *Phenomenology and the Cognitive Sciences, 6*, 527–546.

Colombetti, G. (2011). Enaction, sense-making, and emotion. In J. Stewart, O. Gapenne, & E. Di Paolo (Eds.), *Enaction: Towards a new paradigm of cognitive science*. Cambridge, MA: MIT Press.

Colombetti, G. (2014). *The feeling body: Affective science meets the enactive mind.* Cambridge MA: MIT Press.

Colombetti, G., & Thompson, E. (2008). The feeling body: Toward an enactive approach to emotion. In W. F. Overton, U. Müller, & J. Newman (Eds.), *Body in mind, mind in body: Developmental perspectives on embodiment and consciousness*. New York: Lawrence Erlbaum.

Damasio, A. (1994). *Descartes' error: Emotion, reason, and the human brain.* New York: Avon Books.

De Jaegher, H. (2009). Social understanding through direct perception? Yes, by interaction. *Consciousness and Cognition, 18*, 535–542.

De Jaegher, H., & Di Paolo, E. (2007). Participatory sense-making: An enactive approach to social cognition. *Phenomenology and the Cognitive Sciences, 6*, 485–507.

Evans, D. (2002). The search hypothesis of emotion. *The British Journal for the Philosophy of Science, 53*(4), 497–509.

Furtak, R. A. (2010). Emotion, the bodily, and the cognitive. *Philosophical Explorations, 13*(1), 51–64.

Gallagher, S. (2005). *How the Body Shapes the Mind.* Oxford: Oxford University Press.

Gallagher, S. (2008). Inference or interaction: Social cognition without precursors. *Philosophical Explorations, 11*(3), 163–174.

Gilligan, C. (1982). *In a different voice: Psychological theory and women's development.* Cambridge, MA: Harvard University Press.

Goldie, P. (2004). Emotion, feeling, and knowledge of the world. In R. Solomon (Ed.), *Thinking about feeling: Contemporary philosophers on emotions* (pp. 91–105). New York: Oxford University Press.

Goldman, A. (1995). Empathy, mind, and morals. In M. Davies & T. Stone (Eds.), *Mental simulation* (pp. 185–208). Oxford: Blackwell Publishers.

Greene, J. D., Nystrom, L. E., Engel, C. L., Darley, J. M., & Cohen, J. S. (2004). The neural bases of cognitive conflict and control in moral judgment. *Neuron, 44*, 389–400.

Greene, J. D. (2007). The secret of Kant's soul. In W. Sinnott-Armstrong (Ed.), *Moral Psychology* (Vol. 3, pp. 35–117). Cambridge, MA: MIT Press.

Haidt, J. (2007). The new synthesis in moral psychology. *Science, 316*, 998–1002.

Hatfield, E., Cacioppo, J., & Rapson, R. (1994). *Emotional contagion.* Cambridge: Cambridge University Press.

Jaggar, A. (1989). Love and knowledge: Emotion in feminist epistemology. *Inquiry: An Interdisciplinary Journal of Philosophy, 32*, 151–176.

Jones, K. (2004). Emotional rationality as practical rationality. In C. Calhoun (Ed.), *Setting the Moral Compass: Essays by Women Philosophers* (pp. 333–352). Oxford: Oxford University Press.

Kahneman, D., & Frederick, S. (2002). Representatives revisited: attribute substitution in intuitive judgment. In T. Gilovich, D. Griffin, & D. Kahneman (Eds.), *Heuristics and Biases* (pp. 49–81). New York: Cambridge University Press.

Lewis, M. (2005). Bridging emotion theory and neurobiology through dynamic systems modeling. *Behavioral and Brain Sciences, 28*, 169–194.

Lloyd, G. (1984). *The man of reason: 'Male' and 'female' in Western philosophy.* Minneapolis, MI: University of Minnesota Press.

Lutz, C. (1986). Emotion, thought, and estrangement: Emotion as a cultural category. *Cultural Anthropology, 1*(3), 287–309.

Lutz, C. (2001). Emotion and feminist theories. In J. Mageo (Ed.), *Power and the self.* Cambridge: Cambridge University Press.
Maiese, M. (2013). Moral cognition, affect, and psychopathy. *Philosophical Psychology*. doi:10.1080/09515089.2013.793916
Mumby, D., & Putnam, L. (1992). The politics of emotion: A feminist reading of bounded rationality. *The Academy of Management Review, 17*(3), 465–486.
Orsi, F. (2012). Moral judgment, sensitivity to reasons, and the multi-system view. *The Baltic International Yearbook of Cognition, Logic, and Communication, 7*, 1–19.
Prinz, J. (2004). *Gut reactions: A perceptual theory of emotion*. New York: Oxford University Press.
Ratcliffe, M. (2002). Heidegger's attunement and the neuropsychology of emotion. *Phenomenology and the Cognitive Sciences, 1*(3), 287–312.
Ratcliffe, M. (2007). *Rethinking commonsense psychology: A critique of folk psychology, theory of mind, and simulation*. Hampshire: Palgrave Macmillan.
Ratcliffe, M. (2008). *Feelings of being: Phenomenology, psychiatry, and the sense of reality*. New York: Oxford University Press.
Ridley, A. (1997). Emotion and feeling. *Supplement to the Proceedings of the Aristotelian Society, 71*(1), 163–176.
Simon, H. (1989). Making Management Decisions: The role of intuition and emotion. In W. H. Agor (Ed.), *Intuition in organizations* (pp. 23–39). Newbury Park, CA: Sage.
Schulz, M.L. (2006). *The new feminine brain: Developing your intuitive genius*. New York: Atria Books.
Solomon, R. (1976). *The passions*. Indianapolis: Hackett.
Striano, T., & Reid, V. (2006). Social cognition in the first year. *Trends in Cognitive Science, 10*(10), 471–476.
Thompson, E. (2007). *Mind in life: Biology, phenomenology, and the sciences of the mind*. Cambridge, MA: Belknap Press.
Trevarthen, C. (1979). Communication and cooperation in early infancy: a description of primary intersubjectivity. In M. Bullowa (Ed.), *Before speech: The beginnings of interpersonal communication.* (pp. 321–348). Cambridge: Cambridge University Press.
Whiting, D. (2009). The feeling theory of emotions and the object-directed emotions. *European Journal of Philosophy, 19*(2), 281–303.
Whiting, D. (2012). Are emotions perceptual experiences of value? *Ratio, 25*(1), 93–107.
Zemach, E. (2001). What is emotion? *American Philosophical Quarterly, 38*(2), 197–207.

Chapter 8
Values in the Social Sciences: The Case of Feminist Research

Kristina Rolin

8.1 Introduction

Feminist research is research guided by feminist values (Anderson 2004; Crasnow 2007; Longino 1995; Rolin 2012; Wylie 2007). While many feminist philosophers endorse this motto, their views diverge on the question of what it means for feminist values such as equality and social justice to *guide* research in the social sciences. Some roles granted to moral and social values in research are uncontroversial, whereas some others are less so. Moral and social values are often thought to play an irreducible role in decisions concerning what topics are worthy of research and for what practical purposes scientific knowledge is sought. Also, moral and social values may justify the imposition of constraints on social scientific inquiry by requiring that informants give their informed consent to research and that their privacy is protected. Moving towards a controversial terrain, we find the question of whether moral and social values are allowed to play a role in the decision making processes that scientists are engaged in when they accept something as scientific knowledge, and if they are, what roles they can legitimately play.

Much of the controversy revolves around the ideal of value-free science, the view that moral and social values are *not* allowed to play a role in the decision making processes whereby scientists accept something as scientific knowledge. A number of feminist philosophers argue that the value-free ideal is not feasible, and it should be replaced by another view on the proper role of values in science (Anderson 1995, 2004; Brown 2013a, b; Clough 2011; Crasnow 2014; Fehr 2011; Hawthorne 2010; Intemann 2001, 2005; Kourany 2010; Longino 1990, 1995, 2002; de Melo-Martín and Intemann 2012; Richardson 2010; Rooney 1992; Solomon 2001). While the feminist criticism of the value-free ideal is well received in the philosophy of science community (Biddle 2013; Carrier et al. 2008; Douglas 2000,

K. Rolin (✉)
University of Helsinki, Helsinki, Finland
e-mail: Kristina.Rolin@helsinki.fi

M.C. Amoretti and N. Vassallo (eds.), *Meta-Philosophical Reflection on Feminist Philosophies of Science*, Boston Studies in the Philosophy and History of Science 317, DOI 10.1007/978-3-319-26348-9_8

2007, 2009; Elliott 2011b; Kincaid et al. 2007; Kitcher 2001, 2011; Lacey 1999, 2005, 2013; Machamer and Wolters 2004; Risjord 2007; Root 1993; Steel 2010; Wilholt 2009), it has left us wondering in what ways feminist values can constructively guide research in the social sciences. One model of integrating feminist values with scientific inquiry can be found in Janet Kourany's *Philosophy of Science after Feminism* (2010). According to Kourany, feminist research has paved the way for the ideal of socially responsible science, the view that "sound social values as well as sound epistemic values must control every aspect of the research process, from the choice of research questions to the communication and application of results" (Kourany 2010: 106). While the ideal of socially responsible science succeeds in capturing the feminist commitment to good science in the double sense of goodness—moral and epistemic—it gives rise to further questions: What does it mean to say that sound social values should "control" a research process? Can sound social and epistemic values go hand-in-hand without one undermining the other?

With the aim of shedding light on these questions, I introduce a threefold analysis of how feminist values can legitimately and productively work in social scientific inquiry. I review three well-known arguments against the value-free ideal: an argument based on pluralism with respect to epistemic values (Sect. 8.2); an argument based on inductive risk (Sect. 8.3); and an argument based on value-laden background assumptions (Sect. 8.4). These three arguments are built on slightly different yet overlapping analyses of what it means for a scientist to accept a hypothesis or a theory, and how moral and social values can play a legitimate role in acceptance which is thought to be a core epistemic moment in scientific inquiry. I argue that these three arguments provide us with a framework for understanding how feminist values can legitimately and fruitfully guide research in the social sciences. In Sect. 8.5, I review two normative views which have been proposed as alternatives to the value-free ideal, Miriam Solomon's social empiricism and Helen Longino's critical contextual empiricism. While these two forms of empiricism are put forward as epistemic ideals, they are also political ideals. I argue that they can be interpreted as attempts to implement the liberal democratic values of equality and neutrality in science policy. As epistemic and political ideals they offer justification for the view that some social scientific research can be committed to feminist values explicitly.

My analysis is not meant to be an exhaustive account of the ways feminist values can guide research in the social sciences. For example, *feminist standpoint theory* offers an influential model for thinking about the role of feminist values in the social sciences (see e.g., Crasnow 2009; Harding 2004; Intemann 2010; Rolin 2006, 2009; Wylie 2003, 2012). In this model, feminist values are thought to give rise to collaborative research projects where the process of generating evidence is coupled with a process of empowerment (Rolin 2009: 219). As I think that feminist standpoint theory deserves more space than I am capable of giving to it in this essay, I will postpone an analysis of it to another occasion.

8.2 Plurality of Epistemic Values

When one argues against the value-free ideal, it is not sufficient to show that scientific research sometimes fails to be value-free. One has to show also that the ideal in itself is not feasible—or even if it is feasible under some circumstances, there are reasons which speak against its adoption as a standard of good science. In this and the next two sections, I introduce three arguments aiming to do so. Examples from feminist research are provided to illustrate how feminist values can legitimately interact with empirical evidence and epistemic values in the social sciences.

The first argument is the argument based on pluralism with respect to epistemic values. A number of philosophers argue that the value-free ideal is not attainable because the set of epistemic values includes a variety of criteria and desiderata which cannot be realized at the same time, and non-epistemic values can legitimately play a role in determining which epistemic values scientists emphasize when they evaluate theories (Elliott 2013; Elliott and McKaughan 2014; Kitcher 1993; Kuhn 1977; Longino 1995, 2002; Rooney 1992; Solomon 2001). By epistemic values is meant values which promote the attainment of truth, either intrinsically or extrinsically. As Daniel Steel (2010: 18) explains, an epistemic value is intrinsic when manifesting that value constitutes an attainment of or is necessary for truth, and it is extrinsic when it promotes the attainment of truth without itself being an indicator or a requirement of truth. For example, internal consistency is an intrinsic epistemic value because there is a necessary connection between a theory's statements being true and internally consistent (Douglas 2009: 94). While simplicity cannot be derived from the goal of truth in the same way as internal consistency, it can be argued to be an extrinsic epistemic value on the basis that a preference for a simpler hypothesis promotes truth under some specific circumstances (Steel 2010: 19).

Arguments aiming to undermine the value-free ideal by drawing attention to the plurality of epistemic values follow two strategies. One strategy aims to show that the plurality of epistemic values is a consequence of the plurality of goals in science. Another strategy suggests that the plurality of epistemic values is revealed by studying actual practices of science. The first strategy can be found in Philip Kitcher's (1993) and Miriam Solomon's (2001) works. Kitcher (1993: 94; see also Anderson 1995) argues that the goal of scientific inquiry is more appropriately understood to be *significant* truth than plain truth, and consequently, the goals of scientific inquiry are as many as there are views about significance in science. Moreover, a search for a context-independent conception of significance is misguided because significance reflects the non-epistemic concerns of the age even when such concerns appear to be internal of science (Kitcher 2001: 82). Solomon (2001: 27; see also Elliott 2013) suggests that the goal of scientific inquiry is more appropriately understood to be *empirical success* than truth, and since empirical successes come in many forms, there is plenty of leeway for non-epistemic considerations to steer the direction of research. Both Kitcher (1993) and Solomon (2001) argue that pluralism with respect to epistemic values is epistemically beneficial insofar as it generates and maintains a *distribution of research efforts* in scientific communities among those theories that have some empirical successes. Such a distribution, they argue, is a prerequisite to the long-term epistemic success of science (see also Chang 2012).

The second strategy can be found in the works of Thomas Kuhn (1977) and Helen Longino (1995, 2002) who aim to articulate those values that have actually played a role, explicitly or implicitly, in debates concerning good theories. Whereas Kuhn lists five epistemic values (accuracy, consistency, simplicity, breadth of scope, and fruitfulness), Longino (2002: 185) adds to this list five other values which, she argues, can be called epistemic on equally good grounds: empirical adequacy, novelty, ontological heterogeneity, complexity of interaction, applicability to human needs, and diffusion of power. In Kuhn's view, as well as in Longino's, philosophers should embrace pluralism with respect to epistemic values because it helps them understand how scientists can *rationally disagree* with each other at any given moment (see also Wray 2011: 182). Scientists can do so by interpreting epistemic values in different ways or by giving more weight to some epistemic values than others.

In sum, these arguments are based on an analysis of acceptance as a particular kind of value judgment. In this analysis, acceptance consists of two moments, "valuing" and "evaluation" (McMullin 1983: 5). "Valuing" is about choosing which epistemic values are applied at a particular decision making situation and "evaluation" is about assessing the extent to which a theory realizes the chosen epistemic values. While non-epistemic values are not allowed to play a role in "evaluation," they can legitimately play a role in "valuing" because in this role they are thought to be epistemically harmless or even beneficial if they contribute to an efficient division of research efforts in scientific communities.

I argue that pluralism with respect to epistemic values provides us with a framework for understanding how feminist values can legitimately and effectively guide research in the social sciences. Especially in those areas of scholarship where the object of scientific inquiry is a morally laden phenomenon, moral and social values are needed to guide the interpretation of epistemic values such as empirical adequacy and explanatory power. In such cases a scientist can hardly avoid making moral value judgments because the object of scientific inquiry is already charged with moral values independently of the scientist's judgments. So, even if a scientist chose to ignore the value-rich nature of the object, she would thereby make a moral value judgment ("let us leave the world as we have found it"). When a scientist cannot avoid making moral value judgments, making *morally justifiabl*e value judgments is the morally responsible thing to do.

Ann Cudd's (2005, 2006) theory of oppression is an apt example to illustrate how feminist values guide research in a case where the object of inquiry is saturated with moral and social values. As Cudd (2005: 20–21) explains, defining oppression is a matter of making moral value judgments. According to her definition, oppression occurs if and only if the following four conditions are fulfilled: the harm condition, the social group condition, the privilege condition, and the coercion condition (Cudd 2005: 23, 2006: 25). The *harm condition* requires that there is a harm that comes out of an institutional practice of oppression. The harm may be inflicted by psychological or material forces of oppression. For example, psychological forces may oppress people by reducing their own self-image, by causing emotional stress, or by manipulating their beliefs. Material forces may oppress people by reducing their material resources or by threating them with violence. The *social group condition* requires that the harm is "perpetrated through a social institutions or practice on a social group whose identity exists apart from the oppressive harm" (Cudd 2006: 25). The *privilege condition* requires that there is a social group that benefits from the institutional practice of oppression. The *coercion condition* requires that there is unjustified coercion or force that brings about the harm. Given Cudd's definition, oppression is always *wrong* because its harms are unjustly inflicted on oppressed people (Cudd 2006: 23). As I understand it, this is what makes her definition of oppression *feminist*.

The necessity of working with a morally value-laden definition of oppression has interesting implications for the empirical study of oppression. It is the moral value judgments implicit in the definition of oppression that will enable us to pick out the right cases for the empirical study of oppression. Moreover, theories of oppression should be evaluated not only on the basis of whether they are responsive to empirical evidence but also on the basis of whether they fulfill certain moral and pragmatic criteria. According to Cudd (2005: 36), the moral criterion proposes that "theories of human social phenomena must treat human individuals as morally primary and the account of the normative concept has to be consistent with that assumption." This means that social groups, institutions, and practices should not have a prior moral claim to the individuals who compose them. As Cudd (2005: 21) explains, a theory of oppression should not turn a blind eye to the fact that it is the individuals in the oppressed social groups who suffer from oppression. The moral criterion interacts with the epistemic value of empirical adequacy by suggesting that relevant empirical evidence should include information of the impact that oppression has on individual lives. In addition to the moral criterion, Cudd (2005: 32) proposes an axiological pragmatic criterion suggesting that a good explanation will offer the kind of causal information that is needed to manipulate the phenomenon under investigation. When the axiological pragmatic criterion is applied to evaluate a theory of oppression, it is perfectly legitimate to expect that the theory not only explains how oppression is maintained but also gives advice for ending it.

Cudd (2005) argues that on the basis of the moral and the axiological pragmatic criterion, a structural rational choice theory fares better than some alternative theories of oppression such as the social dominance theory developed in evolutionary psychology. A structural rational choice theory aims to understand how social structures constrain the options individuals have as members of particular social groups, how social structures differently reward or punish individuals on the basis of their social group membership, and how social structures ultimately motivate human behavior by influencing the expected utilities of their choices. The moral criterion is met because from such an analysis we can see how individual lives are influenced by social structures. The axiological pragmatic criterion is met because we can see how individuals participate in the maintenance of oppressive social structures and how such structures can be changed in order to reduce or overcome oppression (Cudd 2005: 46).

While a structural rational choice theory meets the moral and the axiological pragmatic criteria, the social dominance theory fails to do so. According to Cudd (2005: 44), it violates the moral criterion because it equates oppression with dominance and social hierarchy thereby collapsing normative notions into descriptive ones. As a result, the social dominance theory does not offer any tools to distinguish between legitimate and illegitimate forms of dominance and social hierarchy. Also, the social dominance theory does not measure up to the axiological pragmatic criterion because it explains oppression as an outcome of behavioral patterns which are thought to be a function of evolutionary processes. Given the hypothesis that some human beings are evolutionary programmed to dominate, it is hard to see how oppressive institutions and practices could be changed (Cudd 2005: 44).

To summarize, the lesson we can learn from Cudd's theory of oppression is that pluralism with respect to epistemic values provides us with a framework for understanding how feminist values can legitimately guide research in the social sciences. Feminist values can inform the very definition of the object of inquiry thereby guiding the way the epistemic value of empirical adequacy is understood. Also, feminist values can inform the way the epistemic value of explanatory power in interpreted. In the case of oppression, good explanatory theory offers tools for tackling oppressive institutions and practices. Finally, feminist values can be used to evaluate theories alongside with epistemic values by requiring that the moral value of individuals be respected. Next, I will turn to another argument which provides us with a framework for understanding the role of feminist values in the social sciences.

8.3 Inductive Risk Argument

The second argument against the value-free ideal is the inductive risk argument. A number of philosophers argue that the value-free ideal is not feasible because non-epistemic values have a legitimate role to play in the evaluation of risks involved in acceptance (Biddle 2013; Brown 2013b; Douglas 2000, 2007, 2009; Elliott 2011a; Steel 2010, 2013; Wilholt 2009). Whereas the term "inductive risk" is often attributed to Carl Hempel (1965), the most often cited version of the inductive risk argument can be found in Richard Rudner's 1953 article titled "The Scientist qua Scientist Makes Value Judgments." One premise in Rudner's argument is the view that a scientist as scientist accepts or rejects hypotheses, and acceptance involves uncertainty because "no scientific hypothesis is ever completely verified" (Rudner 1953: 2). In accepting a hypothesis a scientist has to decide whether the evidence at hand is sufficiently strong to warrant the acceptance. This decision, Rudner argues, depends on the risks involved. If a scientist accepts a false hypothesis, there may be a cost associated with this type of error. Also, if she rejects a true hypothesis, there may be another cost associated with the other type of error.

The key premise in Rudner's argument is that the assessment of the costs involved in these two mistakes is a matter of moral value judgment (Rudner 1953: 3).

It is important to notice that Rudner's argument builds on a *thick* conception of acceptance. Given this conception, acceptance involves three moments: the assessment of the evidential warrant of a hypothesis, the identification of error-related risks, and a moral value judgment concerning an acceptable level of risk. Rudner's thick conception of acceptance is challenged by Richard Jeffrey who argues that acceptance properly understood is the moment when a scientist assigns a probability to a hypothesis on the basis of evidence (Jeffrey 1956: 237). The two other moments, the identification of risks and their moral assessment, are not part of acceptance proper. In Jeffrey's view, acceptance proper should be distinguished from another conception of acceptance, accepting a hypothesis as a basis for some course of action (Jeffrey 1956: 245; see also Hempel 1981: 394–395; Lacey 2005: 986; McMullin 1983: 8). He suggests also that moral value judgments concerning the importance of avoiding mistakes are not part of what scientists do qua scientists. It is someone else's responsibility to assess the seriousness of risks—or insofar as scientists make such assessments, they do so in some other role, not in their role as scientists (see also Mitchell 2004: 251).

Those philosophers who defend Rudner's thick conception of acceptance argue that it is more relevant to the actual practice of science than Jeffrey's thin conception (Biddle 2013; Douglas 2007; Steel 2013; Steele 2012). In the actual practice of science it is difficult to keep the three moments of acceptance apart because scientists are considered to be experts not only on the evidential warrant of the hypothesis but also on its applicability to policy issues and the potential risks generated by errors (Douglas 2000: 563, 2007: 127–130). Also, while there may be some areas in scientific research where the risk of making errors does not have any moral significance, such areas are becoming rare as most research funding is allocated for applied and policy oriented research (Douglas 2000: 577).

The upshot is that if one endorses Rudner's thick conception of acceptance, then the value-free ideal is not attainable because one moment in acceptance involves non-epistemic values. Heather Douglas (2009) argues that even if the value-free ideal were attainable, it would be morally wrong to live up to such an ideal because it encourages scientists to disregard their moral responsibility. Like most human beings, scientists are morally responsible for their actions and the foreseeable consequences of their actions (Douglas 2009: 67). This means that scientists are morally responsible for the potential harm caused by their making overly strong knowledge claims and downplaying the risk of error (Douglas 2009: 87). When scientists do not make moral value judgments concerning acceptable risk, they can be blamed for being reckless or negligent (Douglas 2009: 81).

The ideal that Douglas proposes as a replacement for the value-free ideal is the ideal of scientific integrity. In her view, scientific integrity consists in keeping non-epistemic values to their proper roles in scientific reasoning, not in keeping them out of scientific reasoning (Douglas 2009: 88, see also 156 and 176). In order to define the proper roles for non-epistemic values, Douglas makes a distinction between a direct and indirect role. Values play a direct role when they act as reasons in themselves to accept a hypothesis or a theory and an indirect role when they act as reasons to accept a certain level of uncertainty (Douglas 2009: 96). While non-epistemic values are not allowed to play a direct role in scientific reasoning, they can legitimately play an indirect role. A direct role is not acceptable because it means that non-epistemic values play the same role as evidence does. As Douglas (2009: 156) explains, "values are not evidence, and should not take the place of evidence in our reasoning." An indirect role, on the other hand, is acceptable because scientists are morally responsible for their knowledge claims and the predictable consequences of making such claims (Douglas 2009: 106).

I argue that the inductive risk argument is useful for feminist philosophers' project of analyzing how feminist values can legitimately guide research in the social sciences. It suggests that especially in those areas of scholarship where the object of scientific inquiry is morally and politically value-laden, evidential standards should be higher than in other areas because of the potentially serious consequences of making mistakes. Let me return to the example I have introduced in the previous section, Ann Cudd's theory of oppression. While Cudd (2005) endorses a structural rational choice theory of oppression, she rejects the social dominance theory on the basis that it does not enjoy sufficiently high evidential warrant given the potentially severe consequences of adopting the theory falsely. The social dominance theory explains the persistence of oppression by suggesting that some social groups have developed a stronger social dominance orientation than others. For example, men's social dominance orientation explains why men tend to dominate women (Cudd 2005: 42). The social dominance orientation of men is explained on evolutionary grounds by proposing that it is an adaptive feature for males insofar as it enables them to maximize sexual access to females (Cudd 2005: 43). There is no corresponding evolutionary incentive for females which explains the asymmetry in men's and women's social dominance orientation. Cudd rejects the theory on the grounds that its empirical support is not as strong as one should expect it to be. The theory is not capable of explaining many features of oppression such as psychological forces of oppression and violence against women (Cudd 2005: 43). Such anomalies are a problem for the theory because accepting the theory in error is likely to have harmful consequences especially for the oppressed social groups. A falsely adopted theory can be used to maintain oppression by justifying oppressive institutions and practices.

Summing up the lessons from the example, the inductive risk argument offers yet another model for understanding how feminist values can legitimately guide research in the social sciences. Feminist values can interact with the epistemic value of empirical adequacy by urging us to adopt higher evidential standards in those cases where the object of inquiry is rich in moral and political values and the stakes in making mistakes are high especially for the disadvantaged social groups. It is important to notice that Cudd is not claiming that the social dominance theory should be rejected because the consequences of accepting the theory may be problematic from a moral point of view. She argues instead that the theory should meet high evidential standards because the consequences of adopting the theory *falsely* are problematic from a moral point of view. Next, I will turn to the third argument against the value-free ideal and its lessons for the philosophy of the social sciences.

8.4 Value-Laden Background Assumptions

The third argument against the value-free ideal is the argument based on value-laden background assumptions. A number of philosophers argue that the value-free ideal is not feasible because non-epistemic values can legitimately influence the choice of background assumptions which play a role in a scientist's decision to accept a hypothesis (Anderson 1995, 2004; Clough 2011; Hawthorne 2010; Intemann 2001, 2005; Longino 1990, 2002; de Melo-Martín and Intemann 2012; Richardson 2010). This argument is based on Longino's contextualist analysis of acceptance which requires some explanation.

In Longino's view, acceptance takes place in a context of background assumptions which are needed to establish the relevance of empirical evidence to a hypothesis or a theory (1990: 43–44, 2002: 127). As Longino (1990: 44) explains, "A state of affairs will only be taken to be evidence that something else is the case in light of some background belief or assumption asserting a connection between the two." While background assumptions may not always "encode" moral and social values, they often do so (Longino 1990: 216). Value-laden background assumptions should not be judged as necessarily "bad" science because it is difficult to see how evidential reasoning could proceed without them (Longino 1990: 128 and 216). Non-epistemic values can influence which background assumptions scientists rely on in their evidential reasoning "without violating constitutive rules of science" (Longino 1990: 83). Whether value-laden background assumptions are acceptable or not will depend on a community practice where they are critically evaluated and either defended, modified, or rejected in response to criticism (Longino 1990: 73–74). As Elizabeth Anderson explains, the value-laden nature of background assumptions is not a problem in and by itself; it becomes a problem only when it gives rise to dogmatism (Anderson 2004: 3). In her view, we need to ensure that "value judgments do not operate to drive inquiry to a predetermined conclusion" (Anderson 2004: 11).

Let me introduce an example to illustrate how value-laden background assumptions can legitimately guide research in the social sciences: a feminist study of divorce. Anderson argues that in the case of Abigail Stewart et al.'s study of divorce (Stewart et al. 1997), feminist values played a role in their research design. While many other studies focused primarily on the negative impact that divorce may have had on spouses and their children, Stewart et al. designed data collection so that it was open to finding evidence of both negative and positive outcomes of divorce (Anderson 2004: 12). In this case, we can find two alternative background assumptions which "encode" moral and social values. One value-laden assumption is that divorce may have mainly negative outcomes (and the main research problem is to understand what exactly the negative outcomes are), and another one is that divorce may have not only negative but also positive (or neutral) outcomes for the affected parties. The feminism in Stewart et al.'s research design was embedded in the latter background assumption.

The value-laden background assumption informing the work of Stewart et al. gave impetus to interesting choices in the research design. First, the team decided to diversify the method of data collection. Besides collecting quantitative data on the well-being of family members, they collected qualitative data on their feelings and subjective interpretation of the events (Anderson 2004: 15). The rationale for doing so was to open the data collection process to the possibility that there may be positive interpretations of the events even in those cases where the divorced parents had financial hardships or behavioral problems with their children. Second, the team adopted a longitudinal study design in order to make room for the possibility that the interpretation of the events can change long after the event of divorce (Anderson 2004: 14). Third, the team paid special attention to the value-laden language that is often used to describe divorce. As a result, they were careful not to describe divorce as a family "breakup" but rather as a family "transformation" (Anderson 2004: 15). Also, they did not to assume at the outset that divorce involves a "trauma" or a "loss," or that it may cause a "life stress" that puts children "at risk" for problems later in life (Anderson 2004: 13).

Anderson's main point is to show that while the research design is guided by feminist values, the values do not prompt the researchers to believe dogmatically that there are positive outcomes of divorce. It is left to the evidence to decide whether there are positive outcomes of divorce. Similarly, the choice of the language does not mean that the researchers try to enforce the conclusion that divorce can be a positive experience. Given the evidence, divorce may turn out to be a trauma for some people and it may turn out to be an opportunity for personal growth and an opportunity to provide children with a peaceful home for some others. Thus, feminist values do not lead the researchers to close the door to the evidence; instead, the values open the door to new kinds of evidence that may come as a surprise to them.

Let me wrap up the three arguments I have presented in this and the previous two sections. The argument based on pluralism with respect to epistemic values constructs acceptance as a value judgment which includes both valuing and evaluation. Whereas valuing is a matter of choosing which epistemic values are important, evaluation is a matter of measuring theories against the chosen epistemic values. Given the pluralism with respect to epistemic values, feminist values can legitimately play a role in the interpretation of certain epistemic values such as empirical adequacy and explanatory power. The argument based on inductive risk suggests that the evaluation of empirical adequacy is a more complex affair than merely determining the degree of evidential warrant. When a scientist evaluates a hypothesis, she is expected to identify error-related risks and to make a moral judgment of their seriousness. Given the inductive risk argument, feminist values can play a legitimate role in drawing attention to the potentially grave consequences that some theories may have in case they are endorsed by mistake. The argument based on value-laden background assumptions reveals yet another dimension in the structure of acceptance. Before a scientist reaches the point where she evaluates a hypothesis or a theory on the basis of evidence, she makes a value judgment concerning the plausibility of particular background assumptions. Feminist values can play a legitimate role in suggesting which background assumptions should inform the research design. Feminist values do not compete with empirical evidence; instead, they can interact with empirical evidence by diversifying it.

This summary brings me to my main thesis. The three arguments against the value-free ideal provide us with a multi-dimensional framework for understanding how feminist values can legitimately and constructively guide research in the social sciences. Also, the arguments give rise to normative approaches which have been proposed as alternatives to the value-free ideal. In the next section, I will assess two approaches from a feminist point of view.

8.5 Successors to the Value-Free Ideal

As Janet Kourany (2010: 57–58) argues, the ideal of value-free science is not merely an epistemic but also a political ideal. It is political in the sense that it plays a central role in the traditional social contract for science where the society is expected to provide scientists with funding (no strings attached) and scientists are expected to provide the society with value-free knowledge (Brown 2009). Given the traditional social contract for science, the value-free ideal seems to be justified by two liberal democratic values: the equality of citizens and the neutrality of the state (Kappel 2014). Whereas the value of equality suggests that citizens should have an equal set of rights and basic powers to influence political decision making, the value of neutrality proposes that the state should stay neutral with respect to more specific conceptions of good life that individuals and social groups can strive to realize within the confines of liberal democratic states. The value-free ideal is thought to do justice to the values of equality and neutrality by insisting that publicly funded scientific research responds to citizens' needs in an inclusive and evenhanded way, and remain neutral vis-à-vis the lifestyle preferences of religious or some other groups in the society (Lacey 2013). Given that the value-free ideal resonates with the political values of equality and neutrality, feminist philosophers are urged to reflect on the question of how these political values can be achieved in case the value-free ideal is not feasible.

In this section I review two ideals which feminist philosophers have proposed as alternatives to the value-free ideal, Solomon's social empiricism and Longino's critical contextual empiricism. I argue that both forms of empiricism can be interpreted as attempts to implement the political values of equality and neutrality in science. Whereas the value-free ideal is an attempt to implement these values at the level of *individual* studies, Solomon's and Longino's social epistemologies are attempts to implement them at the level of scientific *communities*. Solomon's social empiricism suggests that equality and neutrality are best realized by allowing for a diversity of value-laden approaches within scientific communities. Longino's critical contextual empiricism suggests that a community wide rational deliberation is the best way to carry out the political values of equality and neutrality in scientific communities. Let me explain these two ideals in more detail.

In Solomon's (2001) view, non-epistemic values can legitimately play a role in determining which empirical successes a scientist considers most important when she decides to work with a particular scientific theory. Solomon thinks that non-epistemic values can play an epistemically beneficial role in science insofar as they generate and maintain an efficient distribution of research effort amongst those theories that have some empirical successes. For this reason, a normative approach to values in science should not discourage the influence of non-epistemic values at the individual level in determining a scientist's choice of one theory over another (Solomon 2001: 120). For an individual scientist, social empiricism gives only one guideline. An individual scientist should work with empirically successful theories (Solomon 2001: 150–151).

Solomon's main concern is the proper functioning of scientific communities. In her view, an efficient distribution of research efforts requires two things. It demands that each theory receives a fair share of resources depending on its empirical successes, and that non-epistemic values are equally distributed among those theories that have some empirical successes (Solomon 2001: 77, 95, and 117–118). Solomon recommends that science policy makers take steps to cultivate diversity in scientific communities. They should be ready to allocate funding for risky, off-the-mainstream research projects. As they cannot know in advance which research programs will be fruitful, they are better off distributing their bets among several alternative lines of inquiry.

Like Solomon, Longino (1990, 2002) is concerned with the proper functioning of scientific communities. In Longino's view, non-epistemic values can legitimately play a role in a scientist's choice of background assumptions as long as no-one has challenged these assumptions. Individual scientists are not held responsible for policing the role of non-epistemic values in scientific inquiry on their own. Such a responsibility would be too demanding because "there are no formal rules, guidelines, or processes that can guarantee that social values will not permeate evidential relations" (Longino 2002: 50). An individual scientist may not even be aware that her preferred background assumptions resonate with certain non-epistemic values (Longino 1990: 80). For these reasons, a social account of objectivity is needed to make sure that value-laden background assumptions can be identified and criticized.

Whereas Solomon is concerned with the maintenance of diversity, Longino is more concerned with the interactions among scientists from diverse backgrounds. Longino's social account of objectivity is the view that scientific knowledge is objective to the degree that a relevant scientific community satisfies the four norms of public venues, uptake of criticism, shared standards, and the equality of intellectual authority (Longino 1990, 76–81, 2002: 129–131). These four norms are required to guarantee that criticism will actually be effective in bringing about transformation in community members' views (Longino 1990: 76, 2002: 129). In Longino's view, the equality of intellectual authority is needed to disqualify those communities where certain perspectives come to dominate because of the political, social or economic power of their adherents (Longino 1990, 78, 2002: 131). Also, the equality of intellectual authority promotes a diversity of perspectives and a diversity of perspectives is necessary for a critical dialogue (Longino 2002: 131). As Longino (2002: 132) explains, "Not only must potentially dissenting voices not be discounted; they must be cultivated."

Longino's approach differs from Solomon's in that it does not assume that scientific communities are capable of realizing the normative ideal without assigning epistemic responsibilities to individual scientists. The four norms have implications for individual behavior in scientific communities. For example, the norm of uptake means that an individual scientist has an obligation to respond to criticism whereas the norm of shared standards means that an individual critic has an obligation to carry the burden of proof by formulating criticism in accordance with the shared standards of the community. The norm of equality means that each party in the controversy can expect to be taken seriously and treated with respect. Thus, given Longino's normative approach, individual scientists have more epistemic responsibilities than the obligation to work with empirically successful theories.

I argue that despite the differences in Solomon's and Longino's social epistemologies, their views can be interpreted as attempts to implement the political values of equality and neutrality in science. Given Solomon's and Longino's social epistemologies, the values of equality and neutrality can be obtained at the level of scientific communities, and not at the level of individual studies. The value of equality is implicit in Solomon's demand that non-epistemic values be equally distributed among those theories that have some empirical successes (Solomon 2001: 118). It is implicit also in Longino's requirement that community members be treated equally when it comes to their intellectual authority (Longino 2002: 131). As Longino understands it, the equality of intellectual authority is "tempered" in scientific communities because domain specific expertise is not equally distributed among community members. Yet, even those community members who do not have "contributory expertise" in a particular domain (Collins and Evans 2002), can have a reasonable perspective on the domain insofar as they are capable of understanding and assessing some arguments. As Longino (2002: 133, note 19) explains, "Intellectual authority is less a matter of having knowledge than of having cognitive and intellectual skills of observation, synthesis, or analysis, which enable one to make cogent comments about matters concerning which one knows less than another."

Not only is equality present in Solomon's and Longino's social epistemologies; the political value of neutrality plays a role in their emphasis on diversity. Both Solomon's social empiricism and Longino's critical contextual empiricism are designed to allow for a diversity of value-laden approaches in scientific communities insofar as these approaches have some empirical successes. They do not forbid the influence of non-epistemic values on the interpretation of epistemic values or on the choice of background assumptions in evidential reasoning. Instead, Solomon (2001: 150) suggests that those scientists who are in a position to do science policy have a responsibility for the maintenance of diversity in scientific communities. Longino (2002: 134) demands that scientific communities be open to diverse perspectives even when these perspectives come from outside the communities.

My conclusion is that the ideal of value-free science is not the only option when one attempts to implement the liberal democratic values of equality and neutrality in science. Solomon's and Longino's social epistemologies offer an alternative model for realizing these values in science. Given social and critical contextual empiricism, the value of equality is fulfilled insofar as citizens have an equal say in decisions concerning the many value-laden approaches in scientific research. Similarly, the value of neutrality is fulfilled insofar as the diversity of value-laden approaches does justice to the diversity of social values in the larger society. The key idea is that the values of equality and neutrality are realized at the level of scientific communities. Individual scientific studies are not expected to measure up to the ideal of value-free science.

Given that Solomon's and Longino's social epistemologies are consistent with the political values of equality and neutrality, one may wonder how feminist research projects fit into this picture. I argue that they fit in well because the view that equality and neutrality are to be implemented at the level of scientific communities is consistent with the view that some research projects in the social sciences are explicitly committed to feminist values. The very idea that equality and neutrality are best realized in the diversity of value-laden approaches in scientific communities gives a political justification for the pursuit of the so called "advocacy scholarship" such as feminist research in the social sciences (Turner 2009). Also, it is important to keep in mind that feminism does not undermine the basic values of liberal democratic societies; instead, it confirms them by reminding us that equality is still in the making. Consequently, feminist research projects have a legitimate place among the range of value-laden approaches in scientific communities.

8.6 Conclusion

As Sarah Richardson, I believe that the future of feminist philosophy of science lies in the development of "robust and dynamic philosophical frameworks for modeling social values in science" (Richardson 2010: 337). In this chapter I have presented three models for understanding the positive roles that feminist values can play in the social sciences. Given the argument based on pluralism with respect to epistemic values, feminist values can interact with the epistemic values of empirical adequacy and explanatory power so that they determine what kind of evidence is relevant for empirical study and what kind of causal information an explanation is expected to offer. In case the object of inquiry is loaded with moral and social values, scientific concepts can be expected to pick out the right cases for empirical study without neglecting morally and socially significant phenomena. Given the inductive risk argument, feminist values can determine the level of evidential warrant a scientific theory is expected to meet. In case a mistaken theory can have damaging consequences for disadvantaged social groups, we can legitimately expect the theory to meet unusually high standards of evidence. In such cases feminist values advise us to raise the threshold of acceptance. Given the argument based on value-laden background assumptions, feminist values can guide the choice of background assumptions that are needed in evidential reasoning. Value-laden background assumptions can open the inquiry to the discovery of novel empirical evidence.

On the basis of this analysis, I argue that in the social sciences feminist and epistemic values can go hand-in-hand without feminist values undermining epistemic values. Insofar as feminist values interact with epistemic values such as empirical adequacy or explanatory power, they do not compete with epistemic values for the control of inquiry. Instead, they provide an interpretation of these values by specifying what kind of empirical evidence is relevant for scientific inquiry, what kind of causal information an explanation should offer, how high the standards of evidential warrant should be, and what kind of research design can be used to seek for novel evidence. It is important to keep in mind that epistemic values are general criteria and desiderata, and therefore, a context-specific interpretation is often a necessity (Kuhn 1977). As Alison Wylie (2007: 569) explains, "Standards of empirical adequacy are context specific and evolve within distinct research traditions, in response to goals and interests that are external to inquiry as well as to internal theoretical and technical considerations." Given that feminist values do not undermine the epistemic values of empirical adequacy and explanatory power, my analysis is consistent with the view that moral and social values are not allowed to drive the inquiry to a predetermined conclusion (Anderson 2004). While feminist values can legitimately show us the way to evidence, the evidence decides whether a hypothesis or a theory of acceptable.

Acknowledgments I wish to thank Brad Wray for his comments on an earlier version of the manuscript.

References

Anderson, E. (1995). Knowledge, human interests, and objectivity in feminist epistemology. *Philosophical Topics, 23*, 7–58.

Anderson, E. (2004). Uses of value judgments in science: A general argument, with lessons from a case study of feminist research on divorce. *Hypatia, 19*, 1–24.

Biddle, J. (2013). State of the field: Transient underdetermination and values in science. *Studies in History and Philosophy of Science, 44*, 124–133.

Brown, M. B. (2009). *Science in democracy: Expertise, institutions, and representations*. Cambridge, MA: MIT Press.

Brown, M. (2013a). The source and status of values for socially responsible science. *Philosophical Studies, 163*(1), 67–76.

Brown, M. (2013b). Values in science beyond underdetermination and inductive risk. *Philosophy of Science, 80*(5), 829–839.

Carrier, M., Howard, D., & Kourany, J. (Eds.). (2008). *The challenge of the social and the pressure of practice: Science and values revisited*. Pittsburgh: University of Pittsburgh Press.
Chang, H. (2012). *Is water H_2O? Evidence, realism and pluralism*. Dordrecht: Springer.
Clough, S. (2011). Gender and the hygiene hypothesis. *Social Science and Medicine, 72*, 486–493.
Collins, H., & Evans, R. (2002). The third wave of science studies: Studies of expertise and experience. *Social Studies of Science, 32*(2), 235–296.
Crasnow, S. (2007). Feminist anthropology and sociology: Issues for social science. In S. P. Turner & M. W. Risjord (Eds.), *Handbook of the philosophy of science: Philosophy of anthropology and sociology* (pp. 755–789). Amsterdam: Elsevier.
Crasnow, S. (2009). Is standpoint theory a resource for feminist epistemology? Introduction. *Hypatia, 24*(4), 189–192.
Crasnow, S. (2014). Feminist science studies: Reasoning from cases. Paper presented at the FEMMSS conference at the University of Waterloo, August 10–13.
Cudd, A. (2005). How to explain oppression: Criteria of adequacy for normative explanatory theories. *Philosophy of the Social Sciences, 35*(1), 20–49.
Cudd, A. (2006). *Analyzing oppression*. Oxford: Oxford University Press.
Douglas, H. (2000). Inductive risk and values in science. *Philosophy of Science, 67*, 559–579.
Douglas, H. (2007). Rejecting the ideal of value-free science. In H. Kincaid, J. Dupré, & A. Wylie (Eds.), *Value-free science? Ideals and illusions* (pp. 120–139). Oxford: Oxford University Press.
Douglas, H. (2009). *Science, policy, and the value-free ideal*. Pittsburgh: University of Pittsburgh Press.
Elliott, K. (2011a). Direct and indirect roles for values in science. *Philosophy of Science, 78*(2), 303–324.
Elliott, K. (2011b). *Is a little pollution good for you? Incorporating societal values in environmental research*. Oxford: Oxford University Press.
Elliott, K. (2013). Douglas in values: From indirect roles to multiple goals. *Studies in History and Philosophy of Science, 44*, 375–383.
Elliott, K., & McKaughan, D. J. (2014). Nonepistemic values and the multiple goals of science. *Philosophy of Science, 81*(1), 1–21.
Fehr, C. (2011). Feminist philosophy of biology. In E. N. Zalta (Ed.), *The Stanford Encyclopedia of Philosophy*. http://plato.stanford.edu/entries/feminist-philosophy-biology/. Accessed 15 Jan 2015.
Harding, S. (Ed.). (2004). *The feminist standpoint theory reader: Intellectual and political controversies*. London: Routledge.
Hawthorne, S. (2010). Embedding values: How science and society jointly valence a concept—The case of ADHD. *Studies in History and Philosophy of Biological and Biomedical Sciences, 41*, 21–31.
Hempel, C. (1965). Science and human values. *Aspects of scientific explanation* (pp. 81–96). New York: The Free Press.
Hempel, C. (1981). Turns in the evolution of the problem of induction. *Synthese, 46*, 389–404.
Intemann, K. (2001). Science and values: Are value judgments always irrelevant to the justification of scientific claims? *Philosophy of Science, 68*(3), S506–S518.
Intemann, K. (2005). Feminism, underdetermination, and values in science. *Philosophy of Science, 72*(5), 1001–1012.
Intemann, K. (2010). 25 years of feminist empiricism and standpoint theory: Where are we now? *Hypatia, 25*(4), 778–796.
Jeffrey, R. (1956). Valuation and acceptance of scientific hypotheses. *Philosophy of Science, 23*(3), 237–246.
Kappel, K. (2014). The proper role of science in liberal democracy. Paper presented at the conference on the Special Role of Science in Liberal Democracy, University of Copenhagen, November 21–22, 2013.
Kincaid, H., Dupré, J., & Wylie, A. (Eds.). (2007). *Value-free science? Ideals and illusions*. Oxford: Oxford University Press.

Kitcher, P. (1993). *The advancement of science: Science without legend, objectivity without illusions*. Oxford: Oxford University Press.

Kitcher, P. (2001). *Science, truth, and democracy*. Oxford: Oxford University Press.

Kitcher, P. (2011). *Science in a democratic society*. New York: Prometheus Books.

Kourany, J. (2010). *Philosophy of science after feminism*. New York: Oxford University Press.

Kuhn, T. (1977). Objectivity, value judgment, and theory choice. *The essential tension: Selected studies in scientific tradition and change* (pp. 320–339). Chicago: The University of Chicago Press.

Lacey, H. (1999). *Is science value free? Values and scientific understanding*. London: Routledge.

Lacey, H. (2005). On the interplay of the cognitive and the social in scientific practices. *Philosophy of Science, 72*(5), 977–988.

Lacey, H. (2013). Rehabilitating neutrality. *Philosophical Studies, 163*, 77–83.

Longino, H. (1990). *Science as social knowledge*. Princeton: Princeton University Press.

Longino, H. (1995). Gender, politics, and the theoretical virtues. *Synthese, 104*(3), 383–397.

Longino, H. (2002). *The fate of knowledge*. Princeton: Princeton University Press.

Machamer, P., & Wolters, G. (Eds.) (2004). *Science, values, and objectivity*. Pittsburgh: University of Pittsburgh Press.

McMullin, E. (1983). Values in science. *In PSA: Proceedings of the Biennial Meeting of the Philosophy of Science Association, 1982*(2), 3–28.

de Melo-Martín, I., & Intemann, K. (2012). Interpreting evidence: Why values can matter as much as science. *Perspectives in Biology and Medicine, 55*(1), 59–70.

Mitchell, S. (2004). The prescribed and proscribed values in science policy. In P. Machamer & G. Wolters (Eds.), *Science, values, and objectivity* (pp. 245–255). Pittsburgh: University of Pittsburgh Press.

Richardson, S. (2010). Feminist philosophy of science: History, contributions and challenges. *Synthese, 177*, 337–362.

Risjord, M. (2007). Scientific change as political action: Franz Boas and the anthropology of race. *Philosophy of the Social Sciences, 37*(1), 24–45.

Rolin, K. (2006). The bias paradox in feminist standpoint epistemology. *Episteme, 3*(1–2), 125–136.

Rolin, K. (2009). Standpoint theory as a methodology for the study of power relations. *Hypatia, 24*(4), 218–226.

Rolin, K. (2012). Feminist philosophy of economics. In U. Mäki (Ed.), *Handbook of the philosophy of science* (Vol. 13, pp. 199–217). Philosophy of Economics Amsterdam: Elsevier.

Rooney, P. (1992). On values in science: Is the epistemic/non-epistemic distinction useful? *In PSA: Proceedings of the Biennial Meeting of the Philosophy of Science Association, 1992*(1), 13–22.

Root, M. (1993). *Philosophy of social science: The methods, ideals, and politics of social inquiry*. Oxford: Blackwell.

Rudner, R. (1953). The scientist qua scientist makes value judgments. *Philosophy of Science, 20*(1), 1–6.

Solomon, M. (2001). *Social empiricism*. Cambridge: MIT Press.

Steel, D. (2010). Epistemic values and the argument from inductive risk. *Philosophy of Science, 77*(1), 14–34.

Steel, D. (2013). Acceptance, values, and inductive risk. *Philosophy of Science, 80*(5), 818–828.

Steele, K. (2012). The scientist qua policy advisor makes value judgments. *Philosophy of Science, 79*(5), 893–904.

Stewart, A. J., Copeland, A. P., Chester, N. L., Malley, J. E., & Barenbaum, N. B. (1997). *Separating together: How divorce transforms families*. New York: Guilford Press.

Turner, S. (2009). Public sociology and democratic theory. In J. van Bouwel (Ed.), *The social sciences and democracy* (pp. 165–180). Hampshire: Palgrave Macmillan.

Wilholt, T. (2009). Bias and values in scientific research. *Studies in History and Philosophy of Science, 40*, 92–101.

Wray, K. B. (2011). *Kuhn's evolutionary social epistemology*. Cambridge: Cambridge University Press.

Wylie, A. (2003). Why standpoint theory matters: Feminist standpoint theory. In R. Figueroa & S. Harding (Eds.), *Philosophical explorations of science, technology, and diversity* (pp. 26–48). New York: Routledge.

Wylie, A. (2007). The feminism question in science: What does it mean to "do social science as a feminist?". In S. N. Hesse-Biber (Ed.), *Handbook of feminist research: Theory and praxis* (pp. 567–577). Thousand Oaks, CA: Sage.

Wylie, A. (2012). Feminist philosophy of science: Standpoint matters. Presidential Address delivered at the Annual Meeting of the Pacific Division of the American Philosophical Association.

Chapter 9
This is not a Manifesto: Archaeology and Feminism

Pamela L. Geller

9.1 Introduction

Feminism is inescapably political, a qualification that for archaeologists has generated no uncertain amount of ambivalence. Skeptics have long decried that such an association compromises scientific objectivity. This criticism continues to be articulated by both positivistic thinkers (e.g., Flannery 2006) *and* scholars who make gender focal in their studies of the past (e.g., Sørensen 2000; Whitehouse 2007). Conversely, advocates explain that political alignment begets better practice in the present. Reflexivity reveals "epistemic norms," per Wylie (2007), which derail parity and sustain autocratic pedagogy (see also Conkey 2013; Gero 1996; Wylie 1992, 1996, 1997, 2006). Feminist perspectives also deepen archaeologists' discussions of past socio-sexual lives. Rather than begin with the presumption that pre-industrial peoples are reducible to ancient Man and Woman, feminist archaeologists offer evidentiary support for gender identities that are contextual, intersectional, and dynamic throughout life courses (e.g., Battle-Baptiste 2011; Clark 2008; Delle et al. 2000; Geller 2004; Joyce 2000a, b; Meskell 1999, 2000; Voss and Casella 2011). Accordingly, their attention to sex/gender systems in ancient and historic settings illumines the processes that have led to current conditions—the processes of differentiating, the processes of authenticating, or the processes of erasing certain categories of gendered personhood.

Now, some three decades since archaeology's initial engagement with feminist ideas and practices (e.g., Bertelsen et al. 1987; Conkey and Spector 1984), the intellectual descendants of feminist foremothers are themselves in the process of mentoring a new cohort of archaeologists. This fact in and of itself speaks to the discipline's transformation. For those who practice archaeology as feminists—a designation which transcends practitioners' gender—a commitment to the politics

P.L. Geller (✉)
University of Miami, Coral Gables, FL, USA
e-mail: p.geller@miami.edu

M.C. Amoretti and N. Vassallo (eds.), *Meta-Philosophical Reflection on Feminist Philosophies of Science*, Boston Studies in the Philosophy and History of Science 317, DOI 10.1007/978-3-319-26348-9_9

of recognition and inclusion in the present and past has made professional inroads. Yet, there are facets of the discipline that resist change. In this chapter, I consider archaeology's current demographic composition, acceptable research foci, and pedagogical emphases. My focus is North American archaeology. The degree to which this discussion pertains to the practice of archaeology in other national settings is not explored. I do presume that gender inequity is ubiquitous globally though it likely varies in its severity and must be explored contextually.

In explanation of inertia, I explore the connection between these truths and power. One social phenomenon that speaks to this intertwining is the presence of sexism in institutional settings. Numerous archaeologists have documented how modern sexism frames the inferences we draw about ancient cultures (e.g., Conkey 1997; Gero 1993; Gifford-Gonzalez 1993; Kokkinidou and Nikolaidou 2000; Walker 1995; Watson and Kennedy 1991). This epistemological focus is only touched on briefly here. Rather, I treat present-day archaeologists' practices of subtle and obvious sexism. Such reflection makes transparent how violence, whether structural, symbolic, or interpersonal, may seem to be experienced idiosyncratically but in fact is pervasive and exists collectively.

9.2 Demographic Composition: Shifts and Stasis

In a statement approved by its Board of Directors on 26 April 2006, the Society for American Archaeology (SAA) reaffirmed a commitment to promoting diversity in its membership, practice, and audience.[1] Recognition of diversity—"in ethnic origin, national origin, gender, race, age, economic status, lifestyle, physical and/or cognitive abilities, religious beliefs, sexual orientation, work background, family structure, and other perceived differences"—was a hard won gain of earlier civil rights movements. The implication of this public statement is that a continued commitment to diversity will expand archaeology's research queries and concerns, which in turn will generate broader appeal amongst different publics and sustain the discipline's relevance. As the largest organization of professional archaeologists in the Americas, then, how diverse is the SAA?

In answer to this query, it is important to first distinguish between inclusivity and intersectionality. The SAA's aim to diversify and bring Others into the practice and knowledge production of archaeology, I believe, addresses the former. Intersectionality need not follow, however, from inclusivity. As articulated by many feminist scholars, intersectionality refers to lived experiences *cross-cut* by multiple identity variables like gender, race, age, class, and sexuality. The concept has its roots in radical Black feminism, like the variant fomented by the Combahee River Collective. Its members converged on Black nationalism, socialist feminism, and

[1] In 2008, Norder and Rizvi (2008: 14) put forth a list of 12 recommendations designed to foster diversity within the SAA, though the extent to which these have been implemented is unclear.

lesbian feminism. As manifested in their 1977 "A Black Feminist Statement" (Combahee River Collective 1982):

> The most general statement of our politics at the present time would be that we are actively committed to struggling against racial, sexual, heterosexual, and class oppression, and see as our particular task the development of integrated analysis and practice based upon the fact that the major systems of oppression are interlocking.

Subsequent scholar activists extended this intellectual work and political agenda. Yet, identifying and addressing the nuances of discrimination that arise from such complexity, they recognized, presents no small challenge (e.g., Anzaldúa 1987; Crenshaw 1991; Davis 1981; Glenn 1985; Mohanty 1988; Moraga and Anzaldúa 1984; Smith 1983). Hence, while the SAA may aim to include those who have been historically disenfranchised in archaeology (i.e., inclusivity), the organization's neglect of intersectionality presents a limited and dated understanding of how marginalization works. The SAA need not think about oppression as interlocking or messy, I would suggest, because archaeology as practiced in North America remains an undiversified field in many regards. A cursory discussion of the discipline's composition with respect to class, race and ethnicity, and sexuality is illustrative.

Though archaeology has historical ties to upper-class antiquarians, its current composition is primarily middle-class practitioners (McGuire 2008; Trigger 1989). Since the 1970s, however, increasing internal stratification has occurred within academic and Cultural Resource Management (CRM) archaeology. The result is a lower tier of workers without benefits, job stability, or annual pay above the poverty level (Patterson 1999). This lower class of archaeological practitioners likely has a distinct gender dimension, as suggested by broader trends in the academy. As I discuss below, more women than men generally hold part-time (or temporary) and/or non-tenure-track appointments (Curtis and Thornton 2013: 12).

Regarding race and ethnicity, telling are the dearth of archaeologists of color. Granted the SAA has established pertinent committees and tasks forces, such as the Minority and Native American Scholarships Committees, to address skewed representation, but growth proceeds at a glacial pace. In the late-1990s, for example, Franklin (1997: 799) counted four African Americans with Ph.D.s who specialized in anthropological archaeology. This number has increased as indicated by the establishment of the Society of Black Archaeologists in 2011.[2] But, even if quadrupled, it would be statistically insignificant in light of the fact that the SAA boasts over 7000 members. Additionally, from informal documentation, there are 18 living Native Americans who have completed Ph.D.s in archaeology or on issues that bridge sociocultural anthropology and archaeology, and around four students are currently undertaking doctoral research. Given archaeology's historic ties to colonialism in the United States—where African enslaved labor was often used to

[2]The organization currently operates as a listserv and not a formal organization. For more information see www.societyofblackarchaeologists.com/about-us.html.

excavate Native Americans' burial mounds—these sustained inequities appear particularly glaring.[3]

With regard to sexuality, the numbers of LGBTQI members amongst the archaeological ranks go undocumented. While the field still seems to operate under a don't ask, don't tell policy, there are hints of shifting attitudes. In 2014, SAA members who identify as LGBTQI archaeologists and allies convened for the first time to form a Queer Archaeology Interest Group.

Dissimilar from these other aspects of identity, a more measurable shift in gender's demographics has occurred within archaeology. As of 2002, females account for ~40 % of Ph.D.s awarded in the field (Boites et al. 2004) and women archaeologists comprise almost half of the SAA's membership. The historic presence of women in the organization, though always fewer in numbers than their male counterparts, may be one reason for their steadily increasing representation in more recent decades. The SAA was established in 1934 with 25 men and 6 women as signatories of the organization's first constitution (Patterson 1999: 161).

Gender's changing demographic is also likely a consequence of certain archaeologists' engagement with feminist ideas and commitment to political activism from the 1960s onward. Today committees and task forces—namely, the Committee on the Status of Women in Archaeology (COSWA),[4] Task Force on Gender and Rates of Research Grant Submissions, and Task Force on Childcare—work to derail backsliding of hard fought gains. But the metaphoric pipeline leaks as one ascends the hierarchical structure, a point that Margaret Conkey made in 2007, two years prior to her taking up the SAA's presidential gavel. Women, she said "are still grossly under-represented among the faculty of our major Ph.D. institutions" (Conkey 2007: 295). SAA's 2005 Salary Survey offers data in support of her statement. The number of men exceeds women in all positions aside from ones that are temporary or stress non-field "housework," per Gero's (1985) discussion of the sexual division of labor in contemporary archaeological practice. These trends are not unique to the SAA. In 1994, concerned members of the Archaeological Institute of America (AIA) established the Women in Archaeology Interest Group to assess gender equity. A survey of members concluded that gendered divisions of labor and leaky pipelines characterized present-day practices, and feminist-oriented studies of ancient cultures were often marginalized (Allen 2002).

Some time has passed since the dissemination of both the SAA and AIA surveys; follow-up assessments would certainly gauge the extent of more recent shifts in parity and demographics. Seeing that archaeology has not been out of lockstep

---

[3]The dearth of Native Americans earning doctorates is not unique to archaeology. In all academic fields, Patel (2014) reports, they "represented 1.2 % of the U.S. population in 2012 but earned just 0.3 % of the doctorates awarded to U.S. citizens and permanent residents." While discrimination has historically been a factor, low numbers today are often tied to faculty mentorship (or the lack thereof), cultural values, and impoverishment. That poverty impacts Native Americans' decisions, or rather constrains them, speaks to the aforementioned issue of intersectionality's complexity.

[4]COSWA was established in 1974 and involvement has ebbed and flowed in the intervening years (Tomásková 2008).

with broader trends in the academy, however, we may surmise that little change has occurred. According to the American Association of University Professors' "2012–2013 Annual Report on the Economic Status of the Profession," with the exception of those working in baccalaureate colleges, men generally earn higher salaries, and women are more likely than men to hold full-time faculty positions that are non-tenure-track appointments (Curtis and Thornton 2013: 12). Six years earlier another AAUP report offered a hypothetical projection about reaching gender parity; provided hiring and retention of men and women stay equal parity would be achieved in 57 years (West and Curtis 2006: 8). The report's authors, however, did characterize the projection as an "extremely optimistic" one.

9.3 Research Interests: Division of Labor

Of course, there is no guarantee that increased diversity will displace archaeology's hoary topical concerns or the theoretical frames that aid in interpretations about past socio-sexual lives. Rather than paradigmatic shift, inclusivity may engender assimilation of the historically disenfranchised. According to Young (1990: 164),

> Assimilation always implies coming into the game after it is already begun, after the rules and standards have already been set, and having to prove oneself according to those rules and standards. In the assimilationist strategy, the privileged groups implicitly define the standards according to which we all will be measured.

For instance, investigations of sexual division of labor, which both male and female archaeologists have conducted, are suggestive of a deeply entrenched commitment to simplified understandings of gender identities, activities, and relationships in the past. That is, to study sexual or gendered division of labor does not make one a feminist archaeologist. There is a semantic, theoretical, and political difference (Geller 2009b: 72–73; Hegmon 2003: 218–219).

As I have discussed elsewhere (Geller 2009a), in their bioarchaeological studies of labor, investigators surmise that archaeological contextualized skeletal data—overall robusticity, pelvic dimorphism, indices of activity—mark males as producers who labor in economic, political, and/or agricultural realms (e.g., Eshed et al. 2004; Larsen 1997; Maggiano et al. 2008; Marchi et al. 2006; Ruff 2000; Sládek et al. 2007; Standen et al. 1997; Villotte et al. 2010). Discussion of females' identities and activities are minimal. Rather, their reproductive parts, which in these cases are bony pelvises, provide sufficient information about their social identity, woman as reproducer. Interestingly, several of these studies also describe differences between males and females, detect females who exhibit "male"-like characteristics, or essentialize males by eliding their age or social status. Yet, these potential complications to a strict sexual division of labor go unaddressed by researchers. The implicit message conveyed is that the labor of female child bearers is circumscribed to the menstrual or birth hut or family home.

Researchers then disseminate biologically deterministic notions about gender, monolithic representations of men and women, and heteronormative conceptions of family structure. The use of archaeologically contextualized skeletal remains that are ancient and from varied cultural contexts acts to represent such activities and identities as timeless and universal. In these studies, the presumption that a sexual division of labor is a given supersedes discussion about skeletal data as evidence of bio-cultural synergism, individuals' lived experiences, and strategic divergences from socioeconomic norms. The naturalization of the division of labor may in turn be used to rationalize the gender inequity we see play out in current archaeological practice, which I discuss later. As Foucault (1982a: 787) noted, the contemporary division of labor involves the exercise of power inasmuch as it brings into play relations between groups and is characterized by the hierarchization of tasks.

To be clear, I am not arguing that binary gender ideologies or gender divisions of labor never existed in certain past contexts (for a noteworthy example see Sofaer Derevenski 2000). Rather, I take issue with the ideology of separate spheres as indisputable common sense. As Geertz (1983) explained, common sense is both historically constructed and a cultural system. Its analysis has the potential to reveal how culturally illogical or contradictory phenomena are naturalized. While Geertz's examples were drawn from "other" cultures, commonsensical notions held by archaeologists—about bodily difference, gender relations, social roles, etc.—are certainly not exempt from critical analysis. The reiteration of these ideas says more about the patriarchal and heteronormative facets of our modern sex/gender system than the past culture under study (Geller 2009a). That is, when it comes to the sexual division of labor, many archaeologists write a history of the past in the terms of their own present, to borrow Foucault's verbiage (1977: 31). There is a danger in doing so inasmuch as those who investigate and speak with intellectual authority about the past, risk naturalizing socio-sexual relations that are culturally contingent and (pre)historically situated. Here we see how history can ossify into indisputable human nature.

To return to labor, feminist-inspired archaeologists have invalidated such universalizing about sex/gender systems, whether in pre-industrialized or industrialized societies. For instance, Bolger (2013) has identified many Old World cases in which pottery production required multiple stages and cooperation between genders. Archaeologists studying pre-Columbian Maya agriculture have also documented activities involving collaboration, complementarity, and dissolution of a public-private divide (e.g., Hendon 1997; Robin 2002); religio-political performances relied on gender fluidity (e.g., Geller 2005; Hewitt 1999; Joyce 2000a; Looper 2002; Stockett 2005); and much about laborious arrangements changed with European colonization. These examples highlight that labor activities are multi-faceted and in need of contextualization.

Furthermore, I do not regard critical evaluation of the ideology of separate spheres as simply another type of presentism—that is, transporting contemporary feminist agendas and political correctness into the past. Rather, I see it as an important recalibration of non-feminist archaeologists' naturalization of sexual divisions of labor and heteronormative social relations. Inasmuch as feminist archaeologists' empirically based studies can call into question commonsensical thinking about human nature or oversimplifications about pre-industrial societies, their contribution to the overarching feminist movement is a significant one. I also do not draw attention to labor simply because it is a useful heuristic. Labor—as undertaken in the modern sphere, corporate office, global marketplace—has come to dominate many popular feminist discussions about gender performance and parity (e.g., Rosin 2012; Sandberg 2013; Slaughter 2012; Wolf 2013). There is a politicized dimension to studying labor in the past that contemporary archaeological practitioners cannot ignore regardless of how objective and scientific they perceive the field to be. One's work may always be used in the service of unintended ends.

Research interests provide the basis for crafting job descriptions and the outcome of job hires. With that being said, in the 2012–2013 academic year, only 3 of 68 job listings pertaining to archaeology mentioned gender as a desirable research interest—two tenure-track positions and a one-year lectureship.[5] For 2013–2014, 4 of 105 positions, two temporary positions and two tenure-track assistant professor positions, mentioned gender as either a teaching or research interest. Hence, in any given year, around 4 % of all departments are explicitly seeking an archaeologist whose investigative concerns include gender. Rather, some of the research areas mentioned in job descriptions for 2013–2014 included longstanding ones like "technological innovations," "foraging economics," "human ecosystems," and "complex societies." Given the three decades that have passed since archaeologists first fomented a study of gender, many may now concur that it is a significant dimension of identify formation and social organization through space and time. Yet, if current job descriptions are any indication, it is unlikely that recent Ph.D.s will receive financial compensation and symbolic capital in the form of job security from their investigations of gender. Where then is the incentive for graduate students to specialize in gender? Of course, it might be that the research interests of individuals hired bear little resemblance to the original job advertisement, but this is difficult to discern given the lack of transparency in job searches, addressed later in this chapter.

9.4 In and Beyond the Classroom

Seeing that scholars often integrate their research concerns into the courses they teach, the disinclination to hire scholars who study gender could also mean that the topic may go unaddressed in undergraduate classes. One remedy could be to create tenure-lines in the fields of gender and sexuality, which would also serve to validate these topics as viable research interests in the discipline. Teaching on gender in

[5]This information was taken from the Archaeology Academic Jobs Wikis (http://academicjobs.wikia.com/wiki/Archaeology_Jobs_2012-2013) and includes tenure-track assistant professorships, associate or full professorships, non-tenure track positions, and postdoctoral research jobs. Information about the 2012–2014 academic year can be found here: http://academicjobs.wikia.com/wiki/Archaeology_Jobs_2013-2014.

undergraduate classrooms, especially introductory courses, is crucial. An archaeology course can offer evidential support for gender as historically contingent and cultural constructed, thereby interrogating modern common sense understandings that works to naturalize social behaviors and identities. Students may otherwise not acquire a frame for understanding how those norms, which structure their lives in real and immediate ways, can be turned into human nature.

The pervasiveness and subtlety of male privilege in the present is one reason I discuss gender in introductory archaeology courses. I cover this topic at three different junctures throughout the semester: (1) during a lecture on bioarchaeology, when I explain how gender is conceptually and materially distinct from sex (total of 10–20 min); (2) in the context of a larger consideration of social organization and spaces (total 30–40 min); and (3) to ground discussion of Spector's (1993) *What This Awl Means: Feminist Archaeology at a Wahpeton Dakota Village*, a text that I use to teach about marginalization, collaboration, communication, and social identities more generally (total 30–40 min).

As it pertains to social organization, my second area, the more focused topic of household archaeology is instructive. In addition to the standard overview of household archaeology—materials, strategies for their study, functions, layout and stratigraphy, location on the landscape, etc.—I also ask students to interrogate some of the presumptions they may have about houses, such as occupants' relationship to each other (see also Hendon 2006; Romanowicz and Wright 1996). Connections that are neither consanguineal nor conjugal can speak to creation of cohorts by age, ability, occupational specialization, or physiological processes. To illustrate my points, I offer archaeological examples of men's houses and menstrual huts. Students are generally less familiar with the latter than the former given their cursory treatment in or erasure from historic and ethnographic writings. Historians' descriptions of menstrual huts, as Patricia Galloway (1997) has argued, have long represented these places and the physiological processes associated with them (i.e., menstruation, childbirth) in terms of suffering, taboo, pathology, and pollution. Far from areas of banishment, however, I encourage students to deliberate about such female-identified spaces as places of reprieve and ritual necessity by those who sheltered within. As such, we might read agency and empowerment into these spaces, rather than simply biological happenstance or misfortune.

Female-bodied students can empathize with the physiological experience of menstruation and most undergraduates recognize menstrual synchrony given their cohabitation in dorms. The virtual erasure or trivialization of this topic from archaeological research agendas strikes students as anomalous given the ubiquity and frequency[6] of menstruation, as well as the materiality of structures linked to such processes. Menstrual huts are then good to think with because they bring to the fore an unexplored, gendered aspect of many past cultures' social organization and

[6]Of course, I realize that the frequency of menstruation in the present is debatable as underscored by Strassman's (1997) ethnographic study of the Dogon. Her findings, which stress the impact that activity and nutritional stress have on menstrual frequency, are certainly applicable to past peoples, as well.

ritual practices. They remind us that as human beings we are the complex outcome of biological and cultural phenomena. And these spaces indicate just how male privilege continues to plague the field, which is a microcosm for the larger world in which we live.

In total, I spend roughly two class meetings of some 28 (around 7 % of the semester) discussing feminist topics. Yet, nothing I teach throughout the course of an entire semester generates as much discussion during class or in the qualitative comments of my teaching evaluations. For those students who see archaeology as scientific, unbiased, and apolitical, feminism and gender rub them the wrong way. And, evaluations' anonymity makes them fearless. Legal scholar Lazos (2012: 182) has argued that students' harsh judgment of professors who challenge their worldview, their almost visceral reactions, surface in teaching evaluations, and this is but one of many reasons that evaluations should be abandoned or overhauled. (That student teaching evaluations may be holding back women and minorities is another reason.) Seeing that teaching evaluations are often utilized in tenure review, those professors who teach on sensitive topics may be placed at a disadvantage or penalized for course content. Should we not teach on such topics then despite their importance? And not necessarily relevant to gender as taught in the classroom but rather as performed in the classroom, other students have commented that I am "very kind," "pretty," caring, and fashionably dressed. I am also "condescending" and I have a "cold demeanor." These comments suggest that students' views are not just influenced by presumptions about feminism but also culturally appropriate notions about gender norms connected to femininity.

9.5 Power: Subtle and Embodied

In explanation of archaeology's demographic composition and maintenance of acceptable research and teaching concerns, I would like to dwell on power. Of late, one influential social theorist on the subject has been Michel Foucault. Rather than power as phenomena, Foucault (1982a) was invested in understanding its machinations, the relations it produces, the processes surrounding its exercise. "Power," he remarked, "is everywhere; not because it embraces everything, but because it comes from everywhere" (Foucault 1978: 93). Foucault went on to elaborate, "Power is not an institution, and not a structure; neither is it a certain strength we are endowed with; it is the name that one attributes to a complex strategical situation in a particular society" (1978: 93). In brief, he characterized power as domination, not over but throughout. As such, it is omnipresent, rationalizing, and relational.

Bourdieu (1997: 169–171) would augment Foucault's understanding with his recognition of power's exercise as invisible, replicating, involuntary (even by those who are disadvantaged by its effects), and corporeally inscribed. "The effect of symbolic domination (sexual, ethnic, cultural, linguistic, etc.) is exerted not in the pure logic of knowing consciousness but in the obscurity of the dispositions of habitus." Power under these circumstances "is exercised through *rational communication*" (Bourdieu 1997: 83), and as a consequence its arbitrariness is easy to rationalize. Bourdieu has identified this variant of power as "symbolic violence." He continues to explain that inscription of social structures in bodies"—as evidenced by dress, gestures, surgical modifications—produces an "extraordinary inertia" that derails feminists' efforts for political liberation. According to Bourdieu (1997: 172), "raising of consciousness" is not enough to circumvent reproduction of a dominant political order, because sexism becomes inculcated, embodied.[7]

For example, we may deliberate about appropriate archaeological labor, pertinent to my earlier discussion of its sexual division. As feminist archaeologists have recognized, fieldwork and excavation—more so than research that takes place in the laboratory, museum storage room, or archives—is of primacy in the disciplinary culture. Its masculinist associations include notions about romanticized adventure, physical endurance, and hard drinking (e.g., Gero 1985, 1996; Moser 2007). All archaeologists then must embody hegemonic masculinist performances in field settings, regardless of their biological sex, in order to be successful, ideal professionals. Such performances, however, may be more difficult for women to effect given the obstinacy of gender norms. Archaeological practice takes place within encompassing socio-political contexts, and cultural constructions of gender—the researcher's own vs. the one in which fieldwork occurs—might be irreconcilable. When gender is troubled, per Butler (1990), women's physical wellbeing and safety in the field may be jeopardized. Symbolic violence then can manifest as physical violence. Moreover, women may find themselves in a double bind. A personal predilection for femininity may be difficult to reconcile with the public and professional in cultural settings where traditional or restrictive gender norms are appropriate. Of course, these consequences are not just a product of working in social settings where sex/gender systems may differ from the fieldworker's own. As I discuss later, female archaeologists' sexual harassment at the hands of male practitioners still occurs. Thus, a conception of power as subtle and capable of masking domination's arbitrariness or complexity explains how archaeology's practitioners can discourse officially on diversity with the best intentions yet not fully realize equitable demographic, epistemological, or performative shifts.

Psychologists' recent studies of subtle sexism support philosophical meditations about power's unconscious operation. Glick and Fiske (1996, 1997, 2011), whose overarching concern is the ambivalence generated by feminism, have undertaken pivotal work on the subject of subtle sexism. They take their inspiration from racism theorists, though they are aware of some key differences between racism and sexism related to social distance. They do not, however, address the complications wrought from intersectionality. "That a sexist might reject women at work yet fervently embrace them at home," the psychologists write, "cuts to the core of how

---

[7]"[F]eminist theorists who, giving way to habits of thought, expect political liberation to come from the 'raising of consciousness'—ignoring the extraordinary inertia which results from the inscription of social structures in bodies, for lack of a dispositional theory of practices" (Bourdieu 1997: 172).

ambivalence toward women differs from racial ambivalence" (Glick and Fiske 2011: 351). According to Glick and Fiske, sexism today requires more careful assessment for it can masquerade as benevolence. This "paternalistic prejudice" is just as, if not more so, harmful to women given the facility with which it disappears.

A 2003 SAA Member Needs Assessment Survey suggests that benevolent sexism is an entrenched component of archaeology's practice despite discourse to the contrary. The survey queried members about gender inequity with regard to publishing opportunities, funding allotment, symposia invitations, chilly climate, glass ceiling, role models, and family life.[8] From her analysis, Baxter (2005: 8) concluded that "male respondents did not perceive inequities as being common in many key areas of professional participation […] [whereas] female respondents tended to view inequities as common in far greater numbers." Hence, there appears to be a subtle type of sexism at work in archaeology that is not experienced or obvious to all of its practitioners. One striking aspect of Baxter's analysis is data pertaining not to commonness but rarity. When asked the question "In your experience, how common is sexism and chilly climate in the workplace/field," 22 % of female respondents said rare or very rare in contrast to the 52 % said it was common or very common. And in answer to a similar question about glass ceiling syndrome, 13 % of female respondents said rare or very rare in contrast to the 56 % said it was common or very common. These findings indicate that there are also some *female members* of the SAA who do not see gender inequity as a disciplinary problem. Why is that?

Sexism, Glick and Fiske have demonstrated, does not only involve a male who dominates and a female who is subjugated.[9] Women can also endorse BS, their intentional (and appropriate) acronym for benevolent sexism. Power is not exerted by one group over another but rather spreads throughout social relations. As to why, psychologist Virginia Valian, who has written extensively on the subject of gender schemas, notes, "Males and females have similar cognitions about gender and make similar judgments and evaluations of men's and women's behaviors […] cross-nationally, men's and women's attitudes are correlated: nations with more sexist men are also nations with more sexist women" (2005: 200; see also 1998). With this in mind, it is altogether possible that discriminatory practices and beliefs may be held by those who personally purport to be feminists or are sensitive to gender inequities but do not advocate for institutional or structural change.

To extend understanding of subtle sexism as it occurs in archaeology, we can also examine the silencing of discussions about hostile sexism. There is an ironic ambivalence that emerges from five decades of feminists' progressive, structural and ideological transformations. In spring 2013, Kathryn Clancy and colleagues

---

[8]The 2003 survey did not have a question about gender inequity with regard to hiring, whether in academia or commercial contract work.

[9]They do, however, presume that males and females are involved in heterosexual relations; in their words, there is a high degree of "intimate interdependence between the 'in-group' and 'out-group'" (Glick and Fiske 2011: 533). Hence, the heteronormative presumptions at work in their analysis require further attention.

released a study about sexual harassment in biological anthropology (Clancy 2013).[10] More than 20 % of female biological anthropologists who took part in an anonymous survey said that they had experienced "physical sexual harassment or unwanted sexual contact." Often the perpetrator of harassment was a male in an elevated position of power (i.e., field supervisor, committee chair). There are reasons that this number—which is likely higher given underreporting of hostile sexism—may be comparable in archaeology. Both anthropological sub-fields have been slow to diversify, include practitioners who are reticent to embrace political ideologies perceived as compromising scientific objectivity (i.e., feminism), and involve collaborative fieldwork.

Anecdotal evidence suggests that sexual harassment, primarily of women by men, is also a very real dimension of archaeological practice. Yet, the topic remains a non-issue, one of which practitioners rarely if ever speak aloud. The AIA's (1997) Code of Professional Standards briefly mentions discrimination in Section III's Responsibilities to Colleagues: "Professional archaeologists should not practice discrimination or harassment based on sex, religion, age, race, national origin, disability, or sexual orientation; project sponsors should establish the means to eliminate and/or investigate complaints of discrimination or harassment." But, as Wright (2003, 2008) has pointed out, the SAA includes neither gender equity nor sexual harassment in its Principle of Ethics. The situation may be changing, however. As the writing of this chapter neared completion, I received a request from the Southeastern Archaeological Conference (SEAC)[11] to complete a survey on the occurrence of sexual harassment in field situations. The survey may very well reveal that sexual harassment is statistically infrequent in contemporary archaeology. Or, archaeologists may continue to maintain their silence on the subject.

9.6 Speaking Truth to Power?

How then do we speak truth to asymmetrical power if it is subtle and embodied, and even the disadvantaged may be complicit in its perpetuation? Similar to other disciplines, the gender inequities in archaeology are foundational. While structural changes have occurred, hard-won gains can easily be reversed. So that this does not happen, Valian (2005: 209–211) has suggested that academia more broadly implements critical assessment, transparency, and accountability in hiring practices. Most institutions who advertise jobs through the American Anthropological Association's or SAA's online Career Center, the major disciplinary forums for job

[10]In response, AAA released a statement about the organization's zero tolerance policy for sexual harassment. See http://blog.aaanet.org/2013/04/16/zero-tolerance-for-sexual-harassment/.

[11]SEAC, which brings together archaeologists working in the southeastern United States, has a smaller number of members than the SAA. As of spring 2002, the organization was comprised of 1020 members (http://www.southeasternarchaeology.org/about/history/).

seekers, note that they are an Affirmative Action/Equal Opportunity Employer (EOE)[12] and some add that they "encourage" applications from minorities and women. How many departments follow through with this discourse? Since most departments do not communicate their final decisions to applicants, it is difficult to discern who gets archaeology jobs posted. Even the Archaeology Academic Jobs Wiki, which tries to make the job search process more transparent, rarely includes this information.[13] The outcomes of job searches may surprise archaeologists. Or maybe it will not. But either, way critical assessment and greater transparency would highlight the degree to which a department's (or the field's) discourse is consistent with its practice when it comes to cultivating professional diversity. Institutions might then have to account or explain the more glaring deviations. Beyond hiring, transparency and accountability would similarly have to extend to salary allotments, promotions, and allocation of resources.

A greater impediment to gender equity pertains to entrenched gender stereotypes about family and labor. Indeed, even when institutions support family or paternity leave policies that complement those extended to mothers, there seems to be minimal shift in primary caregiving responsibilities. For example, in 1999, the Faculty Committees on Women in Science at MIT released a report documenting pervasive and subtle gender discrimination at the institution. The report's impact reverberated beyond MIT, and similar inequities were identified at other universities and colleges. Happily, changed occurred; "remarkable progress" in the words of those who have authored a 2011 follow-up report (Conrad et al. 2011: 12). Indeed, the follow-up assessment is an excellent example of sustained vigilance so that advances do not retreat and sustained deficiencies are exposed. And both reports highlight how institutions can address discrimination and advocate for change. The recent report verified successful recruitment of women to faculty positions, increases in their salary and lab spaces, and equality in teaching loads, amongst other things. The study, however, documented little shift in faculty's family responsibilities for those with children. Women are still primary caregivers despite the implementation of family leave (which replaced maternity leave). And gendered stereotypes (i.e., "soft and sweet," nurturing, no overt ambition or aggression) remain rigid. In news commentary on the report's findings, the dean of the School of Science at MIT, Marc Kastner, remarked, "The more fundamental issues are societal, and MIT can't solve them on its own" (Zernike 2011). Here is exactly where archaeological research and course content can contribute. Archaeologists may provide crucial insight into myriad past social organizations—distant from our

---

[12]Often this is abbreviated as M/F/H/V, which means minorities, women, people with disabilities, and military veterans. AAA has added information about discrimination with regard to gender identity/expression and sexual orientation/preference. This language does differ contingent on nation, as indicated by postings for positions in Canada.

[13]On the Academic Archaeology Jobs Wiki, 2013–2014, I have seen only one instance where a named person was identified as having received the job position advertised. A debate continues on the wiki about the ethicality of posting the names of those who get jobs. Most respondents still advocate for anonymity.

own in time and space—that can interrogate unsubstantiated common sense about labor's division, domesticity, masculinity and femininity, and childcare.

To further shift perceptions about appropriate gender performances, other archaeologists have stressed the importance of faculty mentoring (e.g., Baxter et al. 2008; Surface-Evans and Jackson 2012). Without downplaying faculty's responsibilities to their students, however, I think it is also important to think about the role of the peer cohort. Demographically speaking, "a cohort may be defined as the aggregate of individuals (within some population definition) who experienced the same event within the same time interval" (Ryder 1965: 845). In anthropology, cohorts are ubiquitous, so much so that the concept often goes uninterrogated. Cohorts, for instance, form the foundation for examining emic perceptions of rites of passage like those associated with girling girls and boying boys—to borrow a phrase from Joyce (2000b)—as well as etic categorizations of generational groupings (i.e., Baby Boomers, Generation X), though it is important to not presume the two are always equivalent (e.g., Ortner 2003, 2006: 83–86). Sociologists, on the other hand, have theorized to a greater degree about the concept itself. Age, or a "birth cohort," is perhaps the most obvious type deployed for understanding the nature, composition, and influence of a group within any given society. But, as sociologist Glenn (2005: 2) has explained, "The events that define cohorts may range from marriage to joining an organization, from entering a graduate program to becoming a parent for the first time."

In anthropology, cohorts are fostered initially through shared pedagogical experiences (and frustrations). Accordingly, interactions between the members of an institutional cohort can make for productive, intellectual collaborations and emotional support that last throughout one's career—in classroom settings, informal contexts, and professional and public venues. But, to be clear, I am not conflating cohort with a collective train of thought. A cohort's varied sociopolitical commitments and lived experiences—cross-cut by age, gender, race, class, religion, etc.—may diverge widely. Such variability also extends to cultural and geographical areas of specialization, the time periods on which individuals concentrate, the intra- and interdisciplinary bridges they build to enrich understanding of humans' conditions, and the contemporary peoples with whom they interact. These all shape archaeological praxis in profound ways that offer important points of departure with regard to lived experiences and intellectual inquiries.

9.7 Conclusion: *Ceci N'est Pas Une Manifeste*

> You don't have to speak truth to power; because they know it already [...] you don't speak truth to anybody, that's too arrogant. What you do is join with people and try to find the truth, so you listen to them and tell them what you think and so on, and you try to encourage people to think for themselves (Chomsky 2010).

I return to the aforementioned MIT assessments. The 1999 and 2011 reports are insightful inasmuch as they highlight the important role that institutions and the individuals that comprise them can play in addressing gender discrimination, as well as other varieties. Implementation of essential policies and ongoing evaluation of the professional climate have ameliorated the more glaring inequities. And yet, as the 2011 report found, there is only so much truth that power is willing and/or unable to hear. Sexism persists and is pervasive because it is intersectional, contextual, and complex—inclusive of structural, physical, and symbolic violences. Chomsky's statements then indicate that hard truths may require creative, collaborative, and calculated alternatives. In this chapter, I have advocated for the interrogation of common sense about social arrangements and identities, an envisioning of educational emphases, and the cultivation of cohort.

To conclude, this chapter is not a manifesto. An assistant professor on the tenure-track, who is versed in power and its production of knowledge, can ill afford to critique institutional structures if she wants to remain employed. *Ceci n'est pas une manifeste*. And yet, in not naming this a manifesto, I echo René Magritte. According to Foucault (1982b: 36), in his homage to the surrealist painting and his work, Magritte names the painting *This is Not a Pipe* "in order to focus attention upon the very act of naming," to call a manifesto to mind. This chapter then does double duty, as a historical and cultural analysis of archaeology, and as a veiled declaration of opinions, beliefs and intentions. The former functions to identify the shifting demographic composition, longstanding reaching concerns, structural and ideological changes born from engagement with feminism, and varieties of sexism in operation. The latter is a strategic effort to realize a more equitable archaeology.

References

Allen, S. H. (2002). The archaeology of the AIA: An introduction. In S. Allen (Ed.), *Excavating our past: Perspectives on the history of archaeological Institute of America* (pp. 1–28). Boston: Archaeological Institute of America.

Anzaldúa, G. (1987). *Borderlands/La frontera: The new mestiza*. San Francisco: Aunt Lute Books.

Archaeological Institute of America. (1997). *Code of professional standards*. http://www.archaeological.org/news/advocacy/132. Accessed May 30, 2014.

Battle-Baptiste, W. (2011). *Black feminist archaeology*. Walnut Creek, CA: AltaMira Press.

Baxter, J. E. (2005). Gendered perceptions of archaeology: A perspective from the SAA member needs assessment survey. *The SAA Archaeological Record, 5*(4), 7–9.

Baxter, J. E., Mayfield, T., O'Gorman, J., Peterson, J., & Stone, T. (2008). Mentoring strategies for women in archaeology: A report of the 2008 COSWA working group. *The SAA Archaeological Record, 8*(4), 15–18.

Bertelsen, R., Næss, J.-R., & Lillehammer, A. (Eds.) (1987). *Were they all men? An examination of sex roles in prehistoric society*. Acts from a workshop held at Ulstein Kloster, Rogaland, Nov. 2–4, 1979. Stavanger, Norway: Arkeologisk Museum i Stavanger.

Boites, S., Geller, P. L., & Patterson, T. (2004). *The growth and changing composition of anthropology: 1966–2002*. http://www.aaanet.org/resources/researchers/upload/Changing-Composition-1966-2002.pdf. Accessed May 27, 2014.

Bolger, D. (2013). Gender, labor, and pottery production in prehistory. In D. Bolger (Ed.), *A companion to gender prehistory* (pp. 161–179). Malden, MA: Wiley-Blackwell.

Bourdieu, P. (1997). *Pascalian meditations*. Stanford, CA: Stanford University Press.

Butler, J. (1990). *Gender trouble: Feminism and the subversion of identity*. New York: Routledge.

Chomsky, N. (2010). Noam Chomsky: speaking of truth and power (interview by D. Tresilian). *Al-Ahram Weekly*, June 3. http://www.chomsky.info/interviews/20100603.htm/. Accessed Aug 20, 2014.

Clancy, K. (2013). "I had no power to say 'that's not okay:'" Reports of harassment and abuse in the field. *Scientific American*. http://blogs.scientificamerican.com/context-and-variation/2013/04/13/safe13-field-site-chilly-climate-and-abuse/. Accessed Oct 30, 2013.

Clark, B. (2008). *On the edge of purgatory: An archaeology of place in Hispanic Colorado*. Lincoln: University of Nebraska Press.

Combahee River Collective. (1982/1977). Combahee river collective statement. In G. Hull, P. Bell-Scott, B. Smith (Eds.), *All the women are white, all the blacks are men, but some of us are brave: Black women's studies* (pp. 13–22). Old Westbury, NY: Feminist Press.

Conkey, M. (1997). Mobilizing ideologies: Paleolithic "art", gender trouble, and thinking about alternatives. In L. Hager (Ed.), *Women in human evolution* (pp. 172–207). London: Routledge.

Conkey, M. (2007). Questioning theory: Is there a gender of theory in archaeology? *Journal of Archaeological Method and Theory, 14*(3), 285–310.

Conkey, M. (2013). The future of gender in prehistoric archaeology. In D. Bolger (Ed.), *A companion to gender prehistory* (pp. 108–120). Malden, MA: Wiley-Blackwell.

Conkey, M., & Spector, J. (1984). Archaeology and the study of gender. *Advances in Archaeological Method and Theory, 7*, 1–38.

Conrad, J., Hopkins, N., Orr-Weaver, T., Potter, M., Rizzoli, P., Sive, H., Staffilani, G., Stubbe, J., Chisholm, S., Gibson, L., Goldwasser, S., Liskov, B., Newman, D., Ross, C., & Samson, L. (2011). *Report on the status of women faculty in the Schools of Science and Engineering at MIT, 2011*. http://web.mit.edu/newsoffice/images/documents/women-report-2011.pdf/. Accessed Mar 21, 2011.

Crenshaw, K. (1991). Mapping the margins: Intersectionality, identity politics, and violence against women of color. *Stanford Law Review, 43*(6), 1241–1279.

Curtis, J., & Thornton, S. (2013). *Here's the new: The annual report on the economic status of the profession 2012–2013*. http://www.aaup.org/file/2012-13Economic-Status-Report.pdf. Accessed Oct 30, 2013.

Davis, A. (1981). *Women, race, and class*. New York: Random House.

Delle, J., Mrozowski, S., & Paynter, R. (Eds.). (2000). *Lines that divide: Historical archaeologies of race, class and gender*. Knoxville: University of Tennessee Press.

Eshed, V., Gopher, A., Galili, E., & Hershkovitz, I. (2004). Musculoskeletal stress markers in Natufian hunter-gatherers and Neolithic farmers in the Levant: The upper limb. *American Journal of Physical Anthropology, 123*(4), 303–315.

Flannery, K. (2006). On the resilience of anthropological archaeology. *Annual Review of Anthropology, 35*, 1–13.

Foucault, M. (1977). *Discipline & punish*. New York: Random House Inc.

Foucault, M. (1978). *The history of sexuality*. New York: Pantheon Books.

Foucault, M. (1982a). The subject and power. *Critical Inquiry, 8*(4), 777–795.

Foucault, M. (1982b). *This is not a pipe*. Berkeley: University of California Press.

Franklin, M. (1997). Power to the people: Sociopolitics and the archaeology of black Americans. *Historical Archaeology, 31*(3), 36–50.

Galloway, P. (1997). Where have all the menstrual huts gone? The invisibility of menstrual seclusion in the late prehistoric Southeast. In: C. Claassen & R. Joyce (Ed.), *Women in prehistory: North America and Mesoamerica* (pp. 47–62). Philadelphia, PA: University of Pennsylvania Press.

Geertz, C. (1983). *Local knowledge*. New York: Basic Books.

Geller, P. L. (2004). *Transforming bodies, transforming identities: A consideration of pre-Columbian Maya corporeal beliefs and practices*. Unpublished Ph.D. thesis, University of Pennsylvania.

Geller, P. L. (2005). Skeletal analysis and theoretical complications. *World Archaeology, 37*(4), 597–609.
Geller, P. L. (2009a). Bodyscapes, biology, and heteronormativity. *American Anthropologist, 111* (4), 504–516.
Geller, P. L. (2009b). Identity and difference: Complicating gender in archaeology. *Annual Review of Anthropology, 38*, 65–81.
Gero, J. (1985). Sociopolitics of archaeology and the woman-at-home ideology. *American Antiquity, 50*, 342–350.
Gero, J. (1993). The social world of prehistoric facts: Gender and power in Paleo-Indian research. In H. du Cros & L. Smith (Eds.), *Women in archaeology: A feminist critique* (pp. 31–40). Canberra: Australian National University.
Gero, J. (1996). Archaeological practice and gendered encounters with field data. In R. Wright (Ed.), *Gender and archaeology* (pp. 251–280). Philadelphia: University of Pennsylvania Press.
Gifford-Gonzalez, D. (1993). You can hide, but you can't run: Representations of women's work in illustrations of Paleolithic life. *Visual Anthropology Review, 9*(1), 3–21.
Glenn, E. N. (1985). Racial ethnic women's labor: The intersection of race, gender and class oppression. *Review of Radical Political Economics, 17*(3), 86–108.
Glenn, N. D. (2005/1977). *Cohort analysis* (2nd ed.). Thousand Oaks, CA: Sage.
Glick, P., & Fiske, S. (1996). The ambivalent sexism inventory: Differentiating hostile and benevolent sexism. *Journal of Personality and Social Psychology, 70*, 491–512.
Glick, P., & Fiske, S. (1997). Hostile and benevolent sexism: Measuring ambivalent sexist attitudes toward women. *Psychology of Women Quarterly, 21*, 119–135.
Glick, P., & Fiske, S. (2011). Ambivalent sexism revisited. *Psychology of Women Quarterly, 35*, 530–535.
Hegmon, M. (2003). Setting theoretical egos aside: Issues and theory in North American archaeology. *American Antiquity, 68*(2), 213–243.
Hendon, J. (1997). Women's work, women's space, and women's status among the Classic-period Maya elite of the Copán valley, Honduras. In C. Claassen & R. Joyce (Eds.), *Women in prehistory: North America and Mesoamerica* (pp. 33–46). Philadelphia: University of Pennsylvania Press.
Hendon, J. (2006). Feminist perspectives and the teaching of archaeology: Implications from the inadvertent ethnography of the classroom. In P. L. Geller & M. Stockett (Eds.), *Feminist anthropology: Past, present, and future* (pp. 129–142). Philadelphia: University of Pennsylvania Press.
Hewitt, E. (1999). What's in a name? Gender, power, and Classic Maya women rulers. *Ancient Mesoamerica, 10*(2), 251–262.
Joyce, R. (2000a). *Gender and power in prehispanic Mesoamerica*. Austin: University of Texas Press.
Joyce, R. (2000b). Girling the girl and boying the boy: The production of adulthood in ancient Mesoamerica. *World Archaeology, 31*(3), 473–483.
Kokkinidou, D., & Nikolaidou, M. (2000). A sexist present, a human-less past: Museum archaeology in Greece. In M. Donald & L. Hurcombe (Eds.), *Gender and material culture in archaeological perspective* (pp. 33–55). London: Palgrave MacMillan.
Larsen, C. S. (1997). *Bioarchaeology: Interpreting behavior from the human skeleton*. Cambridge: Cambridge University Press.
Lazos, S. R. (2012). Are student teaching evaluations holding back women and minorities? The perils of 'doing' gender and race in the classroom. In G. Gutiérrezy Muhs, Y. Niemann, & C. González (Eds.), *Presumed incompetent: The intersections of race and class for women in academia*. Logan: Utah State University Press.
Looper, M. (2002). Women-men (and men-women): Classic Maya rulers and the third gender. In T. Ardren (Ed.), *Ancient Maya women* (pp. 171–202). Walnut Creek, CA: AltaMira.

Maggiano, I., Schultz, M., Kierdorf, H. S., Sosa, T., Maggiano, C., & Tiesler Blos, V. (2008). Cross-sectional analysis of long bones, occupational activities and long-distance trade of the Classic Maya from Xcambó: Archaeological and osteological evidence. *American Journal of Physical Anthropology, 136*(4), 170–177.

Marchi, D., Sparacello, V., Holt, V., & Formicola, V. (2006). Biomechanical approach to the reconstruction of activity patterns in Neolithic western Liguria, Italy. *American Journal of Physical Anthropology, 131*(4), 447–455.

McGuire, R. (2008). *Archaeology as political action*. Berkeley: University of California Press.

Meskell, L. (1999). *Archaeologies of social life: Age, sex, class, et cetera in ancient Egypt*. Malden, MA: Blackwell.

Meskell, L. (2000). Cycles of life and death: Narrative homology and archaeological realities. *World Archaeology, 31*(3), 423–441.

Mohanty, C. T. (1988). Under Western eyes: Feminist scholarship and colonial discourses. *Feminist Review, 30*, 61–88.

Moraga, C., & Anzaldúa, G. (Eds.). (1984). *This bridge called my back: Writings by radical women of color*. New York: Kitchen Table Press.

Moser, S. (2007). On disciplinary culture: Archaeology as fieldwork and its gendered associations. *Journal of Archaeological Method and Theory, 14*(3), 235–263.

Norder, J., & Rizvi, U. (2008). Reassessing the present for an archaeology of the future: Equity, diversity, and change. *The SAA Archaeological Record, 8*(4), 12–14.

Ortner, S. (2003). *New Jersey dreaming: Capital, culture, and the class of '58*. Durham, NC: Duke University Press.

Ortner, S. (2006). *Anthropology and social theory: Culture, power, and the acting subject*. Durham, NC: Duke University Press.

Patel, V. (2014). Why so few American Indians earn Ph.D.'s, and what colleges can do about it. *The Chronicle of Higher Education*, May 27. http://chronicle.com/article/Why-So-Few-American-Indians/146715/. Accessed Jun 2, 2014.

Patterson, T. (1999). The political economy of archaeology in the United States. *Annual Review of Anthropology, 28*, 155–174.

Robin, C. (2002). Outside of houses: The practices of everyday life at Chan N′oohol, Belize. *Journal of Social Archaeology, 2*(2), 245–268.

Romanowicz, J., & Wright, R. (1996). Gendered perspectives in the classroom. In R. Wright (Ed.), *Gender and archaeology* (pp. 199–223). Philadelphia: University of Pennsylvania Press.

Rosin, H. (2012). *The end of men: And the rise of women*. New York: Viking Press.

Ruff, C. (2000). Biomechanical analyses of archaeological human skeletal samples. In M. A. Katzenberg & S. Saunders (Eds.), *Biological anthropology of the human skeleton* (pp. 71–102). New York: Wiley-Liss.

Ryder, N. (1965). The cohort as a concept in the study of social change. *American Sociological Review 30*, 843–861.

Sandberg, S. (2013). *Lean in: Women, work, and the will to lead*. New York: Alfred A Knopf.

Sládek, V., Berner, M., Sosna, D., & Sailer, R. (2007). Human manipulative behavior in the central European late Eneolithic and early bronze age: Humeral bilateral asymmetry. *American Journal of Physical Anthropology, 133*(1), 669–681.

Slaughter, A.-M. (2012). *Why women still can't have it all. The Atlantic, July/August*. http://www.theatlantic.com/magazine/archive/2012/07/why-women-still-cant-have-it-all/309020/. Accessed Oct 2, 2013.

Smith, B. (Ed.). (1983). *Home girls: A black feminist anthology*. New York: Kitchen Table/Women of Color Press.

Society for American Archaeology (2006). *Statement on diversity*. http://www.saa.org/AbouttheSociety/tabid/54/Default.aspx/. Accessed May 27, 2014.

Sofaer Derevenski, J. (2000). Sex differences in activity-related osseous change in the spine and the gendered division of labor at Ensay and Wharram Percy, UK. *American Journal of Physical Anthropology, 111*, 333–354.

Sørensen, M. L. S. (2000). *Gender archaeology*. Cambridge: Polity.

Spector, J. (1993). *What this awl means: Feminist archaeology at a Wahpeton Dakota Village*. St. Paul, MN: Minnesota Historical Society Press.

Standen, V., Arriaza, B., & Santoro, C. (1997). External auditory exostosis in prehistoric Chilean populations: A test of the cold water hypothesis. *American Journal of Physical Anthropology, 103*(1), 119–129.

Stockett, M. (2005). On the importance of difference: Re-envisioning sex and gender in ancient Mesoamerica. *World Archaeology, 37*(4), 566–578.

Strassman, B. (1997). The biology of menstruation in *Homo sapiens*: Total lifetime menses, fecundity, and nonsynchrony in a natural-fertility population. *Current Anthropology, 38*(1), 123–129.

Surface-Evans, S., & Jackson, M. (2012). Feminine voices in archaeology: Promoting community, collaboration, and mentoring. *The SAA Archaeological Record, 12*(1), 22–25.

Tomásková, S. (2008). History of COSWA: Beginnings, ruptures, and continuities. *The SAA Archaeological Record, 8*(4), 8–11.

Trigger, B. (1989). *A history of archaeological thought*. London: University of Cambridge Press.

Valian, V. (1998). *Why so slow: The advancement of women*. Cambridge, MA: MIT Press.

Valian, V. (2005). Beyond gender schemas: Improving the advancement of women in academia. *Hypatia, 20*(3), 198–213.

Villotte, S., Churchill, S. E., Dutour, O. J., & Henry-Gambier, D. (2010). Subsistence activities and the sexual division of labor in the European upper paleolithic and mesolithic: Evidence from upper limb enthesopathies. *Journal of Human Evolution, 59*(1), 35–43.

Voss, B., & Casella, E. (Eds.). (2011). *The archaeology of colonialism*. Cambridge: Cambridge University Press.

Walker, P. (1995). Problems of preservation and sexism in sexing: Some lessons from historical collections for palaeodemographers. In S. Saunders & A. Herring (Eds.), *Grave reflections: portraying the past through cemetery studies* (pp. 31–47). Toronto: Canadian Scholars' Press.

Watson, P. J., & Kennedy, M. (1991). The development of horticulture in the eastern woodlands of North America: Women's role. In J. Gero & M. Conkey (Eds.), *Engendering archaeology: Women and prehistory* (pp. 255–275). Oxford: Basil Blackwell.

West, M. S., & Curtis, J. (2006). *AAUP faculty gender equity indicators 2006*. Washington, DC: American Association of University Professors.

Whitehouse, R. (2007). Gender archaeology and archaeology of women: Do we need both? In S. Hamilton, R. Whitehouse, & K. Wright (Eds.), *Archaeology and women: Ancient and modern* (pp. 27–40). Walnut Creek, CA: AltaMira.

Wolf, A. (2013). *The XX factor: How working women are creating a new society*. London: Profile Books Ltd.

Wright, R. (2003). Gender matters—A question of ethics. In L. Zimmerman, K. Vitelli, & J. Hollowell-Zimmer (Eds.), *Ethical issues in archaeology* (pp. 225–237). Walnut Creek, CA: Altamira Press.

Wright, R. (2008). Sexual harassment and professional ethics. *The SAA Archaeological Record, 8* (4), 27–30.

Wylie, A. (1992). The interplay of evidential constraints and political interests. *American Antiquity, 57*, 15–35.

Wylie, A. (1996). The constitution of archaeological evidence: Gender politics and science. In P. Galison & D. Stump (Eds.), *The disunity of science: Boundaries, contexts, and power* (pp. 311–343). Stanford: Stanford University Press.

Wylie, A. (1997). The engendering of archaeology: Refiguring feminist science studies. *Osiris, 12*, 80–99.

Wylie, A. (2006). Afterword: On waves. In P. L. Geller & M. Stockett (Eds.), *Feminist anthropology: Past, present, and future* (pp. 167–175). Philadelphia: University of Pennsylvania Press.

Wylie, A. (2007). Doing archaeology as a feminist: Introduction. *Journal of Archaeological Method and Theory, 14*(3), 209–216.

Young, I. M. (1990). *Justice and the politics of difference*. Princeton, NJ: Princeton University Press.

Zernike, K. (2011). *Gains, and Drawbacks, for Female Professors. The New York Times*, March 21.

Chapter 10
Measuring the Value of Women: A Feminist Analysis of Economic Categories and Thought

Ruth Hagengruber

10.1 Philosophy, Economics and Feminist Reflection

We have been educated in and are used to cultural adjacencies within which women do not have the same dominant roles and functions as men have. Although the societies in this world are quite varied, they all have in common that women are generally subordinated in most public institutions, albeit to different degrees. Feminist philosophy undertakes the role of studying the reasons and causes of these arrangements. Feminist economic philosophy is a knowledge project that challenges the tradition of economic thought, pointing to its ostracism of women. It investigates why and how economic science has not treated gender issues adequately, for what reason it has denied women's participation and contribution to economic productivity and continues to hinder women to be part of it. The philosophy of feminist economics targets a transformation of economic categories.

In my essay I will investigate what values are at stake in feminist economics and how these relate to the economic principle of self-interested action and market agency. As women are segregated from economic productivity, relegating their proficiency to their care work, we have to ask about the economic accountability of that work and to question how and if this can be put into an economically relevant determination. Further on, some relevant economic categories must be examined, as this is the antipode of house and market, the economic agent and her freedom to act. I will also discuss the perfect market and add some remarks on what the provisional capital is in this context. Actual policies of gender relevance bring us to take up the discussion of whether feminism and neo-liberal economics have the same target, the autonomy of the agent and how this satisfies our issue. With some remarks on

---

R. Hagengruber (✉)
University of Paderborn, Paderborn, Germany
e-mail: ruth.hagengruber@upb.de

M.C. Amoretti and N. Vassallo (eds.), *Meta-Philosophical Reflection on Feminist Philosophies of Science*, Boston Studies in the Philosophy and History of Science 317, DOI 10.1007/978-3-319-26348-9_10

the history of economics I will demonstrate that the path of a gender-just society have been experienced and what women philosophers and economists have added to that. The goal is to improve the scientific approach of economics and the conviction is that this is not possible if gender issues are not part of these new concepts.

10.2 Self-interested Action and Value Definition

The science of economics conceptualizes itself as a value-free and universal methodical approach used to describe self-interested actions by the agents in a market, which is defined as a place for the exchange of goods. The description of this process does not need to include values or evaluative standpoints, however it does not deny that economic agency is restricted and shaped by normative topics. A Jewish society differs from the action parameters of a Moslem society, this from a Buddhist, a Confucian and a Christian one, and so on. Science itself is not bothered by these preconditions. For the different societies it is true that women are not equally present, neither as producers of economic output nor as economic agents. Women are excluded as agents in the market and thus excluded from the relevant positioning of their interests (Strassmann 1997: viii). The sheer fact that women are not visible within the different economic realms is not problematized within the systems themselves. The fact of their absence and the inferior valuation of their contribution to the Gross National Products worldwide is taken to be a result of their lack of participation and a consequence of women's role in society, stereotyping them into activities which are not regarded as being a part of economic productivity.

Nonetheless, the *rational economic agent* is perceived to be gender-neutral (Ferber and Nelson 2003; Strassmann 1993; Justman 1996). The *homo oeconomicus* is a preference-selecting person, an autonomous individual acting in abstract circumstances, "in society without being influenced by society," and acting in an "ideal market" without realizing the ostracisms and segregations that shape this market (Nelson 1995). Although economic science claims that the *homo oeconomicus* is a universal model of economic action, it regards the predicates of autonomy and context-free action capacity as particularly masculine and is convinced that these concepts are justifiably dominant over culturally feminine predicates, like interdependency, interconnectivity and concreteness. Albeit the economic man is predicated as a universal one, it is a particular identity, built upon exploitation and discrimination (Barker and Kuiper 2003: 9).

Cultural organization has placed one at the market place, the other in the house. Often these two kinds of activities are presented as being opposite. When they are designed to be complementary, this is not relevant for the market. The assumption of perfect competition in the market turns out to be irrelevant for the complementary activities of the gendered production endeavors. On the contrary, the market is a male segregation locus, supported by women and caregivers, who provide important labor to support the productivity in the market (Waring 1988: 14). Mainstream economics presents itself as a result of abstract, free and independent individual actions and ignores the fact that its categories depend on the cultural contexts and social norms which comprise a nation, with religion shaping the determining conditions and traits on the paths which economics has developed and has not overcome but confirmed. It ignores the fact that its knowledge belongs to epistemic communities (Strassmann 1993; Longino 1993; Nelson 1993). Preference economics has disregarded important elements that influence the economic process, shape preferences and thus lead to a tautology of self-confirming effectiveness (Hagengruber 2000). Elementary economic principles that should guarantee the transparency of action, equal access to the market, preference for the best, are ignored and even contradicted by this practice. These kinds of social patterns are realized in preference and utility-functions and deepen the traps of cultural backgrounds. Choice theory is built upon tautology and a consequentialist attitude (Dunleavy 1991; Seiz 1999). This way, economic actions perpetuate segregation habits, allowing the maximum benefit from this kind of oppression and segregation (McCloskey 1998). Stating that these kinds of values enter into economic analysis, as every choice is the expression of personal value, the value judgments in all aspects of economics must be considered (van Staveren 2007). There are elementary parts in economic theory which are inherently *qualitative*. Economics, like any science, is referring to social constructions that shape its theory (Nelson 1995). The aim of feminist research on economic concepts is thus to identify these social and cultural constructions that privilege male interpretations of economic issues, identifying where these forms of dominance and subjection tend to devaluate women in the economic realm (Power 2004).

The market is dependent on provisioning activities which are not part of itself. These are gendered social activities, performed mostly in favor of the male of the house and performed by the females. The sex-segregating labor implies the performance of domestic work, care work, subsistence work such as that of the male dependent worker, with unpaid or lower paid labor and even "voluntary" work, often connected to personal and emotional interaction. This social provisional work —although necessary and also acknowledged as such—has presented a challenge to mainstream economics as the traditional image of the economic man is shaped as the autonomous and self-interested individual. Consequently, the thus constituted *homo economics* negated this provisional work as a market relevant action, as neoclassic economics did in demarcation to classic and ancient economics. Although provisional labor is a substantial and necessary part of every market participation and participant, it is excluded as fact-relevant labor and not counted as a relevant input for the *homo oeconomicus* to present himself in the market.

While the male served as a model to explain interaction in the market, in its social delineation the labor that was dedicated to generating these provisional activities was excluded from being awarded market relevance in various regards. Although we know and economists knew that reproductive labor is a necessary part of productive labor, the accountability of these activities was not an elementary part of research (Boserup 1970). This is not what classical economists such as Marx and Engels where occupied with, although they reflected the provisional necessity of this labor and its gender segregating effect. Capitalist endeavors have been targeted towards replacing this labor providing it on the market. Lastly, it has even liberated women from pregnancy. The commodification of nurture and caring provides the opportunity, as does surrogate mothering, to liberate women from domestic accommodation and subordination (Barker and Feiner 2004). Traditional domestic labor is performed by migrant women, who work as maids, nannies, and cleaners on a monetary level of subsistence, suffering the effects of a gendered market which consist of wage gaps, discrimination, the separation of social relations etc. Although in various countries the participation pattern has been transformed, and women have at least partly taken over responsibilities in the areas deemed to be productive work, the changes with regard to the gendered nature of the market have not kept up in order to provide adequate financial rewards (Barker 2005). Although this work is clearly a necessary complement to the productive labor capitalism is dependent on, its essentialism is defined and rewarded in a different way. Domestic work and care occupations are paid less than other occupations requiring the same level of skills. Women continue to bear most of the burden of care and the cost of its production. Caregiving work is not adequately provided for, neither by man nor by the market nor by the state (Beneria 2003).

However, efforts to impute a monetary value to unpaid household labor do exist by either counting the opportunity costs, the replacement costs or defining the costs by using the input-output method (Cloud and Garrett 1996). The opportunity costs method uses the wages a person would earn in the market, the replacement cost method separates the various activities and accounts for their replacement, while the input-output method defines the values added by the household. Studies of unpaid work within households have sought to obtain quantitative measurements of this labor and to increase the attention paid to unpaid work in the designing of new policies (Aslaksen and Koren 2014).

10.3 Categories of Relevance

As women in all realms have endured patriarch-motivated ostracism which has influenced the negation of their rights and access to economic success and well-being, it does not surprise that only 1 % of worldwide property belongs to women. Similar unfavorable figures become evident if we look at the Gross Domestic Product (GDP), which represents the added values of all the production activities of an economic entity, such as a nation or state. All the current calculations of the GDP worldwide demonstrate that women contribute less to it than men do, beside the implicit fact that there is no country in world which does not have a wage gap. Hereby it is interesting to realize that some of the northern countries at some time from the 19th century on recorded women's economic contributions to society. Norway, a country that discontinued the inclusion of domestic work in the GDP in the 20th century in order to adapt to the international standard, had estimated the value of unpaid household work to be about 15 % of the national product in 1943 (Aslaksen 1999). But the problems are not solved by ascribing monetary values to the social labor which is provided by women. Although caring work and gender-segregated labor is held to be essential for the provision of market output, there is a critical answer to the financial inclusion of these activities as a part of and within the consumerist system of the GDP. There is critique to assign this work an analogous status as a consumptive activity. The reason for this is that from the point of view of social provision, such domestic and caregiving activities are partly even anti-consumptive and thus contradictory to the accounting system of the GDP. For example, food preparation at home reflects an anti-consumerist attitude when compared to the more GDP relevant consummation of (fast) food provided in the market. This anti-consumptive attitude becomes even more evident in educational commitments. It is a primary issue to educate children to renounce excessive consumption. Women are "piecing together" goods which would otherwise be wasted. Hazel Kyrk, an early economist introduced the home economics of waste and labeled '*waste*' as a productive resource (Van Velzen 2003). She was a pioneer in the field of the consumer- economics philosophy of waste prevention. This is only a limited example of some of the ideas that are connected to female concepts of economics and it is evident that this kind of economics is fairly different than what is statistically included in the GDP. This kind of waste economics as an example of household economics is different to what happens when waste, destruction and even crime are regarded to be a countable part of it: "It counts napalm and counts nuclear warheads and armored cars for the police to fight the riots in our cities," as Robert Kennedy explained in his speech at the University of Kansas in 1968 (see Kennedy 1968).

Women's work, as far as it addresses this kind of social caregiving labor, is assigned as being even counterproductive for national economic growth. Indeed, these activities are not part of it, although national growth is effectively stimulated —paradoxically—by what seems at first glance to be an anti-consumptive action. This is not what classical economists such as Marx and Engels propagated. When they finally took into account and reflected on the necessary cost of reproducing the labor force, they did not understand its relevance also as an anti-consumptive attitude and they did not understand the gendered nature of these kinds of actions. This concept has to be reinvented from a feminist perspective, as these characteristics have been hitherto neglected (Picchio 1992). Intentionally anti-consumptive behavior can be distinguished within these traditionally gendered activities, which leads to the reduction of the gross domestic product. But this kind of action is undeniable and undoubted and also a socially rewarded activity which consequently, although not directly, supports well-being, both in the social and the economic sense.

Nowadays a variety of different models to determine the factors of growth have been developed, which also take the gender aspect into account. Gross domestic product models which point to education as a determinant have been conceptualized, others investigate fertility as a variable key issue influencing economic growth, all factors of the above mentioned gender aspects. Analyses of gender and development (GAD), women and development, and women in development (WID) have benefited from that and have become important fields in the analysis of macroeconomic objectives. It has now become apparent that these factors have always affected trade and finance (Çağatay et al. 2007; Beneria 2003). This kind of research has changed the international policies of the World Bank and the United Nations and allows a clearer analysis on the relationship between world poverty and gender topics. The effects of macroeconomic policies of structural adjustment and the liberalization of global trade and finance have been looked at from a feminist point of view. A change in (global) policies has taken place. Recommendations to government budgets according to their effects on gender equity form a part of national politics and demonstrate that these fields of action, which were unrightfully discriminated against by mainstream economics, are an important field of economic growth (Walter 1999: 425).

From the macro perspective we will turn again to the micro perspective and the feminist critique in the field. Women claim their participation in economics and the labor market, as well as the acknowledgment of caregiving labor and domestic work. But, although always excluded officially from access to productive institutions, there is no doubt that at all times and in all places in the world women have effectively contributed to economic productivity. This is true for agricultural societies as well as for industrial ones, and beyond organized markets. However, it can only be indirectly proved that they have provided endowments for the groups they belonged to and consequently, as dependents of these, also for themselves. We must also commend that at all times and everywhere women were have been sought out as specialized laborers in various crafts during earlier periods as well as for the industrial and service economies. Today we have been informed that the participation of women in labor raises investments in the market. The policies of Western countries enforce the participation of women in the labor force with the aim of supporting the recovery of economies after downturns (Çağatay and Ertürk 1995). Women have acted within the economic system and were always requested to do so, but at the price of inferior evaluation compared to their male companions. The monetary rewards of labor are gendered up until today. This happens not only because access to the market proves to be gender-segregated. The most favorable position of the white well-educated married man who is the beneficiary of the care and housework of a wife, is as evident as is the most disadvantaged position of the single mother, who not only earns less in the market, is less educated but provides the care work which is *not* economically rewarded. This does not result from the market as a formal institution, however, but results from the reshaping of conditions that are crucial to performing in the market and which are not of service to women, or possibly only for the group of women who is able to engage other women for this domestic work, and who then face the problems of underpayment and work exploitation instead. How this exclusive political situation was established and "scientifically" required, is demonstrated in the reflections of neo-classic labor market theorists, who were by no means "objective," much more, they were intentionally gender segregating. Women's wages should be pegged at a presumably "fair" market-determined level even when their productivity warrants a higher wage, as the neoclassic theorist Pigou proposed (Pujol 2003: 30). However, if male wages fall below the value of their marginal product, Pigou recommended state intervention to raise these wages to their efficiency level. Independent from the fact that women's labor was demanded from industrial employers particularly in specific female industries, they were intentionally segregated from the market, and characterized as unskilled and low grade workers. Raising female productivity was defeated exemplarily, as this could cause a "debacle of industry," or worse, men's rights to jobs and their status as breadwinners might be challenged (Pujol 2003: 31; Taylor-Mill 1868). Proletarian anti-feminism as well as neo-classic theory have contributed to female segregation. Female activities in the labor market even became criminalized when females were seeking for work to prevent the starvation of their children. Equal pay for women was opposed and the plea was for minimal wages, in order to hinder women's access to industrial training. Marshall confirmed that women were not economic agents, as—consequently—neither their own income, nor their situation of dependency and meager household budgets allowed them to act on their own. They only could maximize the utility of the household (Pujol 2003; Forget 1999).

Albeit domestic labor is not counted as market provisioning activity in neoclassical economics, strong assumptions about women's roles have determined economic theory and shaped the gendered nature of economics. Marriage and family were implicit presumptions of economic reflection, thus subjecting women to predominant expectations about their duties and making them objects of exclusion (Folbre 1983; Folbre and Hartmann 1988). Women's activities are explicitly excluded as non-market activities (Warning 1988). Women cannot choose an array of options, but do best to choose between motherhood and family options. Thus, women are not able to act freely and rationally to their own advantage, and therefore are not part of the capitalist marketplace, much more they are options in the marketplace of men (Kuiper 2001).

We observe the interest of economic theorists in maintaining the essentialist roles for women in the house and this is an element of ancient and of modern economics. For Aristotle the hierarchical functions of the two sexes constitute his metaphysical and scientific outline and hereby also the categories of economic organization. The male is by nature superior, and the female inferior, one rules and the other is ruled; this principle of necessity extends to all mankind, he says in his Politics (Pol 1254 b 13–15). While the superior species reigns by reason, the subordinate provides the necessary goods by its natural functions and capacities, which are not rational, but corporeal and tangible. Women's lives are related to fertility and nourishment. Aristotle claims that everyone benefits from this kind of organization, as all participants proceed according to their *ergon* and their virtue. This teleological concept motivates the entire community to attain what is thought to be good. Women are an instrumental part of this, ruled by the male, whose mandate on the family does not command obedience to the law "but to his will" (Fememias 1994: 168).

From this ancient tradition, two main arguments still play a role as critical counterparts of today's feminist critique. In Aristotle, the housefather is dependent on the performance of the house, which allows him to be independent and equal within the community of equals, that is it gives him the honor of an independent citizen who judges according to his own well-being and is not forced to follow others. Economic independence in the Aristotelian political concept is the condition sine qua non that enables citizenship in the best sense. The metaphoric use of the *pater familias* who becomes the role model of the political leader reflects how the original economic power and function sharing conquers the public sphere. Economics was vital in reproducing patriarchy and stabilizing it, also in assigning its functional reasoning to the public domain. We observe thus that the function based order of the house as an economic entity not only was a necessary preposition that enabled the house father to take part in public life, it also became a model for the public in general. The consequence of this organizational transmission was that the provisional activities which were ascribed to the female in the house led to her ostracism in public. The once established subordination of the female as part of the house alienated her from public life and reduced the development of her skills to the provisioning of non-market capacities. Hannah Arendt pointed to the inadequacy of this transmission and criticized the confusion between political reasoning and domestic arrangements (Pitkin 1994). After the breakdown of the Ancient economic theories, which were criticized even during the Renaissance, social norms and political institutions were established yet again on the basis of this ostracism and now branded the development of the economic theory as a market place where women had no part. Classic economics emphasize the necessity of social provisioning in its regard as an enabling capacity for participation in the labor market. It was aware of the provisional and substantial functions of the original concept of home economics, as is the case in Marx and Engels and others, but it was not successful in ending the segregation of women from the market. Neoclassical economics reasoned about its difference to home economics while deepening the arguments of gender segregation (Poujol 2003; Finley 1973). Women's functions as part of the economic process did not change. They did not gain the right to perform themselves, as Engels stated: "To emancipate woman and make her the equal of the man […] remains an impossibility as long as the woman is shut out from social productive labor and restricted to private domestic labor" (Engels 1972: 221). All institutions, the political, the judicial and, as we have seen, also the economic institutions have insisted after the French Revolution and beyond, on keeping women within the circumstances dictated by the ancient theory which dominated hundreds of years before and which was part of a system, that was claimed to be abandoned. The voices of women, however, have become louder since then and have also started to articulate their intentions in the economic realm. Olympe de Gouges, Mary Wollstonecraft and others such as Harriett Taylor Mill and Charlotte Perkins Gilman followed, criticizing the concepts of male dominated and traditional economics (Halldenius 2014; Taylor-Mill 1868; van Staveren 2003).

The definition of the transcendent good of Ancient philosophy had became individualized in the epoch of the rationalism and classic economic theory and women started to claim their own rights at this time. The determinant role of the house father, who had dominated the agricultural tradition turned into the dominant breadwinner of the industrial society. He replaced the conceptual design of the former. He was challenged by his participation in public life, directed the economics of the family and had, like the former, the wife as his persistent provisionary. The Aristotelian concept branded the social norms over centuries and was even present when economics started to distance itself from house economics, as it had already branded the cultural paths. But the dominance of the Aristotelian model was not undisputed and various attempts were made throughout the centuries to establish equal participation of women in society. Aristotle had been a powerful opponent of his teacher Plato, who criticized a society that segregated women from the good of the republic, locating them in the dark of the *oikos*, where reproduction and the maintenance of bodily needs determined their lives. According to Plato, there is no profession which should exclusively be fulfilled by man or woman, because of their sex, as this would lead into nonsense. If we proceeded like that, we would soon have to decide if a cobbler should be hairy or not, concludes Plato ironically in his Republic (Republic 455 d 6–11). According to the platonic proposition "many women are better than many men at many things" (Republic 455 d 3–4) Plato claimed the same rights for both sexes and challenged women to strive for the public good. Platonic theory espoused various concepts of how women were liberated from domestic and care work and how children's education became a public affair. However, the platonic concept was fairly criticized by both males and females, as Plato concern was not to give women autonomy, he abolished the nuclear family, which Annas (1976: 307) criticized to establish the state as an "impersonal institution of a 1984 type".

Difference and equality concepts drove the interests of ancient economic philosophy and are still at stake today. Thus, not only the tracks of Aristotelianism have shaped the economic theory up till now, also Platonic models have been revived. Allusions to it characterize classic economics, such as the theories of Marx and Engels as well as the Utopian literature of the Renaissance, which was an early period of the woman's movement. Platonic inspired philosophy implied far reaching concepts on the change of social norms and economic action, as such revived in the concept of Thomas Morus *Utopia* and in the *Sun City* of Tommaso Campanella, who both—beyond all differences of an egalitarian versus a meritocratic model—presented an economic model based on gender equality (Hagengruber 2015).

10.4 Controversies Over Autonomy

Autonomy, however, is not a goal shared by all feminist theorists, but regarded as a problematic and double edged argument. Platonic thought and those models emerging from it which develop an equality approach which concedes to cutting off gender specific commitments is depreciatingly commented by difference feminism and interpretations that controvert proposals that separate women from their social relations (Kabeer 2002: 28). Current critique has been uttered by Frazer, stating the amiability between feminism and neoliberalism. In her view, the feminist claim for autonomy and the right of individuality has given new significance to neoliberal politics, assigning a moral meaning to market economics and thus supporting the key ingredients of neoliberalism. Frazer warns that the neoliberal intention is to tap a reservist army of women, employing them in low-paid jobs (Frazer 2013; Funk 2013). Other feminists defend what they hold to be liberal values against communitarian romantics. Equality values, so Nussbaum, do not hurt emotions of care and morals. It must be questioned whose interest is being served by casting women's autonomy as a horror picture of a solitary, egoistic, self-sufficient being. Humans are and remain separate individuals "who always continue to have their separate brains and voices and stomachs, however much they love one another," asking if self-sufficiency is such a bad thing for women to pursue, alluding to the activities of Rebecca West in *The Egoist* and the practiced tabooing and segregation of this standpoint in feminist circles (Nussbaum 2000: 238; Rollyson 2009; Goetz 2007). There is a good deal of excitement about women who dare to plead for the values of individuality and selfishness, as in the case of Rand, who offends all expectations of feminine behavior in her individualist philosophy. She propagated the "immorality of altruism," explaining the exploitation of women as immoral, as: "altruism means not merely kindness but self-sacrifice [...] That you live for others [...] is immoral" (Rand and Branden 1964).

It is a feminist philosophic concern to reflect women's contributions to economics in general and to criticize women's segregation and ostracism from economic activity and the goods that can be obtained by it. It criticizes economic hierarchies and privileges based on gender prejudices and resulting from the subordination of women (Barker and Kuiper 2003: 1). Inequalities stem from social norms which determine discrimination and political subjection that hinders women in obtaining equal access to the market. Gender driven interests have supported the unjust distribution of resources and opportunities. Scientific methods were constructed based on gendered social norms and have deepened the paths of female ostracism as opposed to increasing economic well-being.

We have learned that the market structure is not the pure, formal, and non-corruptive institution which allows access to all those who want to participate. Competition and pricing depend on pre-market provisioning activities which shape the transactions before the market takes effect. Patriarchy exercises an influential market power. There is no reason to claim perfect competition in which the participants are assumed to have the same market power, when cultural norms disadvantage the female group as a whole in doing so. Women cannot exert a market force as they cannot apply themselves as self-interested individuals. On the contrary, we see that economic theorists have intentionally contributed to deepening the social segregation following these social norms by developing scientific instruments to prevent females from access to the market and to competition. Patriarchal norms shaped scientific theorems that actively hindered women in offering their capacities in supply and demand structures in different markets. The neoclassical description of labor markets in which occupations are chosen freely by individuals acting alone and of their own free will, is a self-contradicting institution, preventing the participation of those who otherwise provide relatively decisive advantages for those who are allowed to take part in the market.

The capitalist market is not a free market for everyone, but is reshaped by social provisioning, which is in itself not part of its competition, but results from gender exploitation. Women, who are engaged in the provision of domestic labor and social provisioning are not able to act freely and rationally to their own advantage and therefore are not part of the capitalist marketplace, on the contrary, they are options in the marketplace of men. *Self-interested action* as a relevant theorem of capitalist economics has not been executed in its genuine sense. On the one hand, it is mixed up with advantages provided contingent to social norms and their gender segregation implications. On the other hand, self-interested action is not possible for those who are the hindered by these exact gender-segregation creating norms. While one side benefits, conditions for the others deteriorate. There is no transparency in the access to the market. Today, however, we know that gender disparity not only harms women and girls but also "comes at a great cost to people's well-being and to countries' abilities to grow sustainably, to govern effectively, and thus to reduce poverty" (Barker and Kuiper 2003: 13). Living in an unjust way is not only harmful for those who are discriminated against, but for everyone. This at least has been brought to our attention by feminist economics.

References

Annas, J. (1976). Plato's 'republic' and feminism. *Philosophy, 51*(197), 307–321.

Aslaksen, J. (1999). Gross domestic product. In J. Peterson & M. Lewis (Eds.), *The Elgar companion to feminist economics* (pp. 411–418). Cheltenham, Northampton: Edward Elgar.

Aslaksen, J., & Koren, C. (2014). Reflections on unpaid household work, economic growth, and consumption possibilities. In M. Bjørnholt & A. McKay (Eds.), *Counting on Marilyn waring: New advances in feminist economics*. Bradford: Demeter Press/Brunswick Books.

Barker, D. K. (2005). Beyond women and economics: Rereading "Women's work". *Signs, 30*(4), 2189–2209.

Barker, D. K., & Feiner, S. F. (2004). *Liberating economics. Feminist perspectives on families, work, and globalization*. Ann Arbor: University of Michigan Press.

Barker, D. K., & Kuiper, E. (2003). Introduction: Sketching the contours of a feminist philosophy of economics. In D. K. Barker & E. Kuiper (Eds.), *Toward a feminist philosophy of economics* (pp. 1–18). London: Routledge.

Benería, L. (2003a). *Gender, development, and globalization. Economics as if all people mattered*. New York: Routledge.

Boserup, E. (1970). *Woman's role in economic development*. New York: St. Martin's Press.

Çağatay, N., & Ertürk, K. (1995). Macroeconomic consequences of cyclical and secular changes in feminization: An experiment at gendered macromodeling. *World Development, 23*(11), 1969–1977.

Çağatay, N., Elson, D., & Grown, C. (Eds.). (2007). *Feminist economics of trade*. New York: Routledge.

Cloud, K., & Garrett, N. (1996). A modest proposal for inclusion of women's household human capital production in analysis of structural transformation. *Feminist Economics, 2*(3), 93–119.

Dunleavy, P. (1991). *Democracy, bureaucracy and public choice: Economic models in political science*. London: Pearson.

Engels, F. (1972). *The origin of the family, private property and the state*. New York: International Publishers.

Femenías, M. L. (1994). Women and natural hierarchy in Aristotle. *Hypatia, 9*(1), 164–172.
Ferber, M. A., & Nelson, J. A. (Eds.). (2003). *Feminist economics today. Beyond economic man*. Chicago: University of Chicago Press.
Finlay, M. (1973). *The ancient economy*. Berkley: University of California Press.
Folbre, N. (1983). Of patriarchy born: The political economy of fertility decisions. *Feminist Studies, 9*(2), 261–284.
Folbre, N., & Hartmann, H. (1988). The rhetoric of self-interest: Ideology of gender in economic theory. In A. Klamer, D. N. McCloskey, & R. Solow (Eds.), *The consequences of economic rhetoric* (pp. 184–203). Cambridge: Cambridge University Press.
Forget, E. L. (1999). *The social economics of Jean-Baptist say. Markets and virtue*. London: Routledge.
Frazer, N. (2013). Feminism resurgent? Confronting the capitalist crisis in the Neoliberal era. *Fortunes of feminism: From state-mages capitalism to neoliberal crisis* (pp. 189–243). London: Verso.
Funk, N. (2013). Contra Fraser on feminism and neoliberalism. *Hypatia, 28*(1), 179–196.
Goetz, A. M. (2007). Gender justice, citizenship and entitlements: Core concepts, central debates and new directions for research. In M. Mukhopadhyay & N. Singh (Eds.), *Gender, justice, citizenship and development* (pp. 15–58). Ottawa: International Development Research Centre.
Hagengruber, R. (2000). *Nutzen und Allgemeinheit*. St. Augustin: Academia.
Hagengruber, R. (2015). Cutting through the veil of ignorance. Rewriting the history of philosophy. *The Monist, 98*, 1–19.
Halldenius, L. (2014). Mary Wollstonecraft's feminist critique of property: On becoming a thief from principle. *Hypatia, 29*(4), 942–957.
Justman, S. (1996). *The autonomous male of Adam Smith*. Norman, London: University of Oklahoma Press.
Kaaber, N. (2002). *Citizenship and the boundaries of the acknowledged community: Identity, affiliation and exclusion*. IDS Working Paper 171. Brighton: Institute of Development Studies
Kennedy R. F. (1968). Remarks at the University of Kansas, March 18, http://www.jfklibrary.org/Research/Research-Aids/Ready-Reference/RFK-Speeches/Remarks-of-Robert-F-Kennedy-at-the-University-of-Kansas-March-18-1968.aspx/. Accessed January 3, 2015
Kuiper, E. (2001). *The most valuable of all capital. A gender reading of economic texts*. Amsterdam: Thela Thesis
Longino, H. (1993). Subjects, power and knowledge: Description and prescription in feminist philosophy of science. In L. Alcoff & E. Potter (Eds.), *Feminist epistemologies*. London: Routledge.
McCloskey, D. (1998). *The rhetoric of economics* (2nd ed.). Madison: University of Wisconsin Press.
Nelson, J. A. (1993). The study of choice or the study of provisioning? Gender and the definition of economics. In M. A. Ferber & J. A. Nelson (Eds.), *Beyond economic men: Feminist theory and economics* (pp. 23–36). Chicago: University of Chicago Press.
Nelson, J. A. (1995). Feminism and economics. *The Journal of Economic Perspectives, 9*(2), 131–148.
Nussbaum, M. C. (2000). *Women and human development. The capabilities approach*. Cambridge: Cambridge University Press.
Picchio, A. (1992). *Social Reproduction: The Political Economy of the Labour Market*. Cambridge: Cambridge University Press.
Pitkin, H. F. (1994). Conformism, housekeeping, and the attack of the blob: The origins of Hannah Arendt's concept of the social. In B. Honig (Ed.), *Feminist interpretations of Hannah Arendt*. University Park: Pennsylvania State University Press.
Power, M. (2004). Social provisioning as a starting point for feminist economics. *Feminist Economics, 10*(3), 3–19.
Pujol, M. A. (2003). Into the margin! In D. K. Barker & E. Kuiper (Eds.), *Toward a feminist philosophy of economics* (pp. 21–37). London: Routledge.

Rand, A., & Branden, N. (1964). *The virtue of selfishness. Centennial edition mass market*. New York: Paperback.

Rollyson, C. (2009). *Rebecca West: A modern Sibyl*. Bloomington: Universe.

Seiz, J. A., & Janet, A. (1999). Game theory and bargaining models. In J. Peterson & M. Lewis (Eds.), *The Elgar companion to feminist economics* (pp. 379–390). Cheltenham: Edward Elgar.

Strassmann, D. (1993). Not a free market: The rhetoric of disciplinary authority in economics. In M. A. Ferber & J. A. Nelson (Eds.), *Beyond economic men: Feminist theory and economics* (pp. 54–68). Chicago: University of Chicago Press.

Strassmann, D. (1997). Editorial: Expanding the methodological boundaries of economics. *Feminist Economics, 3*(2), vii–ix

Taylor-Mill, H. (1868). *Enfranchisement of women*. London: Trübner.

van Staveren, I. (2003). Feminist fiction and feminist economics. Charlotte Perkins Gilman on efficiency. In D. K. Barker & E. Kuiper (Eds.), *Toward a feminist philosophy of economics* (pp. 38–55). London: Routledge

van Staveren, I. (2007). Beyond utilitarianism and deontology: Ethics in economics. *Review of Political Economy, 19*, 21–35.

Van Velzen, S. (2003). Hazel Kyrk and the ethics of consumption. In D. K. Barker & E. Kuiper (Eds.), *Toward a feminist philosophy of economics* (pp. 38–55). London: Routledge.

Walters, B. (1999). Growth theory (macro models). In J. Peterson & M. Lewis (Eds.), *The Elgar companion to feminist economics* (pp. 419–426). Cheltenham: Edward Elgar.

Waring, M., & Steinem, G. (1988). *If women counted. A new feminist economics*. San Francisco: Harper & Row.

Chapter 11
The Woman of Reason: On the Re-appropriation of Rationality and the Enjoyment of Philosophy

Pieranna Garavaso

> *The content of women's great science, rather, is human kind, and among humanity, men. Her philosophy is not to reason, but to sense.*
>
> Immanuel Kant, *Observations of the Feeling of the Beautiful and Sublime*

11.1 Introduction

I tailored my title after a famous work by Genevieve Lloyd, *The Man of Reason "Male" and "Female" in Western Philosophy*, first published in 1984,[1] in which Lloyd argues that Cartesian dualism has been coopted in Western philosophy and used to link maleness or masculinity with mind, reason, rationality, activity, and culture, while femaleness or femininity[2] has been connected with body, emotion,

[1]In quoting Lloyd, most feminist philosophers take her to be charging Descartes of being in some way responsible for *both* the crystallization of mind/body dualism and its broadly socio-political and cultural consequences with regard to the power differential between men and women. From my cursory reading of Lloyd's important book, I believe that the second charge is not as well supported as the first. This is my reason for using the term "coopted." Another feminist philosopher, Karen Green, also used "the woman of reason" as the title of her book *The Woman of Reason: Feminism, Humanism and Political Thought* (1995).

[2]De Beauvoir (1949), among others, stressed the distinction between sex and gender by insightfully claiming that "no one is born a woman." Later feminist theorists such as Judith Butler (1990) have pointed out that sex like gender is at least in part socially constructed and thereby cannot be treated like a natural kind. These are valuable insights and I will for that reason use male/female or masculine/feminine oppositions as equivalently powerful symbolical tools.

P. Garavaso (✉)
University of Minnesota, Morris, MN, USA
e-mail: garavapf@morris.umn.edu

M.C. Amoretti and N. Vassallo (eds.), *Meta-Philosophical Reflection on Feminist Philosophies of Science*, Boston Studies in the Philosophy and History of Science 317, DOI 10.1007/978-3-319-26348-9_11

irrationality, passivity, nature, etc.[3] The denunciation of these dualisms and of their socially oppressive power in Western literature, history, culture, and philosophy is a common theme in feminist literature[4] and Lloyd's book has become canonical in the denunciation of the variety and breadth of the above mentioned dichotomies. It is clear from this list of dichotomies that feminist philosophers have seen important connections between many forms of social discrimination that may seem *prima facie* totally unrelated, such as the symbolically charged oppositions between emotion and reason and animal and human. This is an important accomplishment on the part of feminist analyses of all of the above core notions of Western cultures.[5] I do not call into doubt the pervasiveness of these symbolic pairings: a cursory look at histories of philosophy, literature, arts, even science textbooks will provide ample evidence of the influence of these dualisms.[6]

Raia Prokhovnik does not merely reassert Lloyd's central thesis that "the maleness of the Man of Reason [...] is not [a] superficial linguistic bias. It lies deep in our philosophical tradition" (Lloyd 1984: xviii); she also makes two other points that will be central to my discussion in this paper. First, these dualisms are so entrenched in the language and cultural fabric of many Western civilizations to carry with them hierarchical distinctions almost without any conscious realization on the part of the speakers in these cultures: "Dichotomies such as reason/emotion and man/woman represent fundamental polarities, fixed deep within Western philosophy and reflected in the structures of our languages. The two polarities also represent two expressions of hierarchical power relations expressed in social practices in patriarchal society" (Prokhovnik 1999: 1). The existence of unconscious biases or gender schemas (see Haslanger 2008) leads to a "naturalization" of difference and power imbalance; this result is dangerous: no one is really responsible for social discrimination if the power differential is regarded as natural, unavoidable, and unchangeable.

[3]Plumwood (1993) distinguishes between "dualism" and "dichotomy," ascribing only to the former the implicit assumption of a power or value hierarchical imbalance between the two sides; Prokhovnik (1999) does the opposite, ascribing only to "dichotomy" a political relevance. I use these terms as neutral, indicating a distinction which does not necessarily embed a social discrimination.

[4]See Plumwood (1993: 442–445), for a more recent overview of the dualisms that are widely mentioned in feminist literature and for an extended discussion of the notion of dualism as a politically charged distinction.

[5]For a sober list of outrageous statements on women, see Plumwood (1993: 444, Footnote 5). Warren (1990) lists three types of connections between women and nature: symbolic, pragmatic, theoretical; it may be interesting to explore whether there are the same types of links between women and logic or argumentation.

[6]For philosophical sources, see Agonito (1977). Three classical discussion of gender bias in science are Bleier (1984), Fausto-Sterling (1985), Hubbard (1990). For an engaging discussion of gender bias in scientific textbooks, see Martin (1991). For a recent listing of misogynistic texts, see Morgan (1989).

Second, Prokhovnik stresses the oddness encapsulated in the title of this essay: the friction between being a woman or being female and being rational or possessing reason:

> According to the logic of dichotomous thinking in the Western cultural tradition the term "rational woman" is a disjunction resting on a confusion of categories. The idea of "rational woman" is, in terms of our inherited and still vital cultural assumptions, a paradox. But these cultural assumptions, such as the priority given to mind over body, upon which the self-contradictory nature of the term "rational woman" rests, are not necessary features of the logic of concepts (Prokhovnik 1999: 1).

Prokhovnik insightfully points out that this friction resembles a contradiction and as such it seems to indicate a logical or conceptual conflict. In a different passage, mentioning Moira Gatens' work, Prokhovnik further states, "other writers have noted that there are also latent *conceptual* connections in the dominant Western cultural tradition which can be explored between reason, masculinity, truth and intellect on the one hand, and sense, femininity, error and emotion on the other (Gatens 1991: 94–95)" (Prokhovnik 1999: 2). The focus of my discussion is the thesis briefly stated in the final sentences of Prokhovnik's quote, i.e., the denial of the conceptual incompatibility of reason and female/feminine.

In this paper, after some preliminary comments on the moral and metaphysical significance of the attribution of rationality, I will focus on feminist work on logic, namely Andrea Nye's condemnation of logic and Val Plumwood's criticism of the standard notion of negation in classical logic. As a matter of fact, not much work has been done in this area and yet it is very much needed; moreover, it is worth reflecting on why logic and mathematics[7] have proven so resistant to feminist critiques. I propose to look closely at some of Gottlob Frege's reflections on negation in one of his later *Logical Investigations*. It will appear clear, I hope, that Frege's notion of negation is not easily pegged in the general category of "Otherness" that Plumwood uses to characterize negation in classical logic. In the second half of the paper, I discuss the claim that the adversarial style of argumentation in philosophy, especially analytic philosophy, is hostile to feminist goals and perhaps greatly responsible for the low numbers of women engaged in academic philosophy. Against this hypothesis, I claim that a more naturalistic perspective on logic can avoid essentialism and provide a feminist friendly and pluralist view of logic, human reasoning, and philosophical argumentation.

[7]When I started working on this essay, I strove to find all discussions of mathematics from a feminist approach and I only found interesting and hopefully influential discussion of pedagogical issues surrounding the teaching of mathematics (and sciences) to boys and girls. This literature was mostly produced in the 90s and focused in educational settings. I was not able to find much theoretical writing on mathematics and feminist philosophies.

11.2 The Significance of the Alleged Opposition Between Women and Rationality

Why is a discussion of the relationship between conceptions of rationality and reason[8] and feminist philosophies important? In her introduction to *The Man of Reason,* Lloyd (1984: ix) states,

> Reason has figured in western culture not only in the assessment of beliefs, but also in the assessment of character. [...] Past philosophical reflection on what is distinctive about human life, and on what should be the priorities of a well-lived life, has issued in character ideals centered on the idea of Reason.

What is at stake is probably the most important feature of humanness; "rational" is a normative term that bestows the possession of one of the core features of personhood together with self-consciousness, the capacity for moral responsibility or moral relevance, and the use of a language[9]. The weightiness of rationality for personhood makes the rhetorical separation and exclusion of women, femaleness, and femininity from reason quite troubling. Denying that women possess the same level of rationality as men possess and then claiming that rationality is a necessary condition for personhood, as most supporters of psychological definitions of personhood do,[10] is equivalent to relegating women either to not being full-fledged persons or, if we accept degrees of personhood, to being only partial persons. I doubt that any contemporary metaphysician would take seriously the thesis that female human beings are persons to a lesser degree than their male counterparts; hence a reexamination of the notion of rationality is in order.

Some more recent feminist studies spur exactly this reexamination and propose more inclusive notions of rationality. Prokhovnik (1999: 6) describes this goal as follows:

> One of the major conclusions of the current work will be that, rather than throwing out "reason" altogether as a "male" construction, the most conceptually plausible way for feminists to move forward from the exclusion of women from rationality, and the identification of women with irrationality, is to recognize that the notion of reason has been very narrowly defined and can be extended to take into account its immanent element of emotion.

Prokhovnik focuses on the overcoming of the alleged conflict between emotion and reason as a crucial step toward creating a notion of rationality according to which both men and women can meaningfully be claimed to be rational.

---

[8]I use reason and rationality as interchangeable, although I assume that it is possible to distinguish between them. For the sake of my discussion, both have been contrasted to emotion, nature, passivity, etc., which have consistently been associated with femaleness and femininity.

[9]See for example Daniel Dennett's six necessary conditions for personhood discussed in his (1976).

[10]See John Locke's *Essay Concerning Human Understanding* (1975), Sects. 9 and 26, Chap. 27, Book II. For contemporary psychological approaches to personal identity, see Quinton (1962), Noonan (2003), Shoemaker (1970).

Deborah Heikes (2010, 2012) proposes a new feminist notion of "virtue rationality" that incorporates the philosophical reflections of Ludwig Wittgenstein, Robert Nozick, and Robert Audi. Heikes's virtue rationality stresses the normative element that I pointed out supports the strong connection between rationality and personhood. For Heikes, rationality is neither a biological feature of some or all human beings nor is it a merely mental or psychological characteristic; in arguing that rationality is a virtue, she stresses its ethical relevance and its character as an ideal, perhaps never fully achieved but always worth striving for. There is hope that with the development of these more inclusive notions of rationality, the expression "the woman of reason" will stop sounding any more unusual or oxymoronic than the expression "the man of reason" does.

These newly developed perspectives on more inclusive notions of rationality are promising and research in these areas must be encouraged. However, in developing these new directions, it may still be useful to raise the question of what is wrong with the previous notion of reason or rationality that we are attempting to revise. Not until that question is asked and answered precisely, can an appropriate expansion be made. Given the breadth of this question, I will look at one area that has been the focus of some feminist criticism: the use of negation in classical logic.[11]

11.3 Classical Logic as One Major Target of Feminist Critiques

In *Words of Power A Feminist Reading of the History of Logic* (1990) Andrea Nye claims that there is some innate conflict between women and logic,[12] especially in the final chapter of this book where she ultimately dismisses the allegedly "simplest" explanation of the exclusion of women from philosophy, the church, power, etc. based on the political imbalance between men and women in favor of the suggestion that women end up distorting or ignoring their experiences by embracing logic (Nye 1990: 176–177). Ayim (1995: 801) summarizes Andrea Nye's controversial condemnation of classical logic as follows:

[11]Marjorie Hass (2002) provides a detailed outline of Luce Irigaray's influential critique of classical logic; I discuss Plumwood's essay because she focuses on negation, points out its alleged "otherness," and defends this view in a readily accessible fashion.

[12]The alleged friction between women and logic or rationality is sometimes mentioned to explain the low number of women working in analytic philosophy as practiced in America and in other English speaking parts of the world. While the paucity of women in the STEM fields has generated much useful research and work, the limited number of women in philosophy has not generated much debate among professional philosophers until quite recently. For a few exceptions, see Van Camp (2006), Haslanger (2008), Antony (2012), Paxton et al. (2012), Hutchinson and Jenkins (2013), and Tyson (2014).

> [According to Nye] the logical enterprise is inherently self-contradictory for feminist thinkers. Feminists who attempt to use logic to demonstrate its shortcomings are doomed to failure; arguing against logical claims is self-defeating, for the critic will be sucked into the maelstrom of logical tradition. Hydra-headed, the logical monster will thrive rather than perish under the sharpened edge of argument; hence the critic succeeds only in strengthening the very endeavour whose shortcomings she attempts to expose. This is guaranteed to happen, according to Nye, because in entering the debate, one is thereby committed to the terms of the debate. The critic herself will be devoured in the process, for logic was constructed to eliminate the voices and concerns of women. "The feminist logician speaks from a script in which the master always wins."

In Ayim's insightful representation of Nye's position, logic and feminism are at opposite sides: a feminist cannot be a logician because logic is the product of men's power over and oppression of women. Nye goes as far as suggesting that women have to give up taking into account their own experiences to embrace logical reasoning. Yet, this statement is clearly false as any teacher of logic may testify; Ayim herself answers in the affirmative the question which she raises in the subtitle of her essay "Can a feminist teach logic?" Thus, is there anything to salvage and to learn from Nye's feminist approach to logic? More recent reflections on the links between logic and feminist philosophies provide viable approaches.

Plumwood discusses the possibility of a feminist logic in her "The Politics of Reason: Toward a Feminist Logic," published in 1993. Plumwood (1993: 437) reasserts that "the historical construction of reason as masculine has structured its dominant forms not only in an exclusive and oppositional relation to women, but to the characteristics and areas of life that they have been taken to represent, such as emotionality, bodiliness, animality, and particularity." Because of these indicted characteristics, logic is characterized not only as rational, of course, but also as abstract, universal, and normative. Logic is meant to be the way we ought to think; women do not seem by nature to think logically; hence, women are not rational. This argument is invalid; there seems to be little purpose in attempting to improve on it. Although Plumwood implicitly accepts that the above argument underlies much discussion on reason/rationality and women, she advances the discussion by providing a more focused discussion of the link between logic and feminist philosophy. Let us thus look at this narrower topic.

Plumwood's main focus is classical logic, which we can assume to include propositional and predicate logic as well as Aristotelian logic, i.e., the areas usually taught in an introductory course in logic in most Western educational institutions. In the section of her article devoted to the specific discussion of logic, Plumwood develops one main charge against classical logic: classical logic is regarded as the Logic, while there are alternative logical systems such as relevant or paraconsistent logics that do not present some of the undesirable features that characterize classical logic. One such feature is the fact that negation can be taken as representing an exclusive type of "otherness" or total opposition between A and not-A. Plumwood argues that the negation of classical propositional logic as well as the relation between a term and its complement in Aristotelian logic—such as the opposition between "dogs," i.e., the set whose members are all and only dogs, and "non-dogs," i.e., the remainder of the universe where there is all that is not a dog, e.g., a squirrel, a proton, a chair and Pieranna—because of their stark opposition, cannot capture more nuanced forms of negation:

> Classical logic provides an account of otherness which has key features of dualistic otherness. The negation of classical logic is a specific concept of negation which forces us to consider otherness in terms of a single universe consisting of everything. In classical logic, negation, ($\sim$p), is interpreted as the universe without p, everything in the universe other than what p covers, as represented in the usual Venn diagram representing p as a figure surrounded by a square which represents the universe, with $\sim$p as the difference. […] [W] hat is important for the issue we are considering here is that $\sim$p can then not be independently or positively identified, but is entirely dependent on p for its specification. Not-p has no independent role, but is introduced as merely alien to the primary notion. […] This corresponds to […] a logic of presence and absence in which the other is specified as the absence of the condition specified by p, rather than as an independent other. […] In the phallic drama of this p-centered account, there is really only one actor, p, and $\sim$p is merely a receptacle (Plumwood 1993: 454).

Plumwood spells out more in detail her symbolic reading of the Venn diagram figure in which the circle symbolizing p is surrounded by an undifferentiated non-p that is homogenized, "offers no resistance" to the penetrating of p, and "is indistinguishable from the rest of the universe." This is probably the most colorful reading of the relation between an Aristotelian class and its complement class; yet for anyone who has ever taught classical logic to hundreds of students, it is hard not to resist Plumwood's picture and claim instead that p and $\sim$p are indeed co-dependent, that using conversion, obversion, and contraposition quickly takes away the temptation to think that not-p is any more irrelevant than p, that the class of "students who are parents" is any more nor any less important or dominant than the class of "students who are not parents" since each of them may be symbolized with "p" or "$\sim$p." Plumwood seems to be suggesting that when we call one area non-p as opposed to p, then we make the former dependent upon the latter in the sense that the former is specified only in relation to the latter. But this view of dependency, even if granted, would concern only the expression, i.e., the signifier, and would not automatically apply to what is symbolized by "p," or the signified. To understand non-p requires a prior understanding of p, but that relation of dependence does not translate into the two areas signified by p and non-p—as the above examples attest. Of course, this may be just one unconvincing critique and there may be a lot more punch in a feminist analysis of the limits of negation in classical logic. The disagreement between Classical and Intuitionist understanding of the negation of p which leads in Intuitionistic mathematics to the rejection of the principle of Excluded Middle is a well-known example of the fact that the negation of classical logic is not as universally accepted as Plumwood's discussion suggests.

One final point concerning Plumwood's criticism of negation in classical logic: I assume that the main point of this criticism may be taken to be that this—let's call it "exclusionary"—notion of negation is not the only notion of negation or otherness that human beings find useful and conducive to rational thinking or reasoning. There may also be the belief that this exclusionary notion of opposition is related to the dualisms mentioned at the beginning, where the two sides of a distinction, such as reason and nature or body and mind are understood as radically opposite to each other, without allowance for varying degrees of difference or more nuanced forms of opposition. But if this is the worry that leads to a form of rejection of the negation of classical logic, then clearly the solution is not that of throwing away the baby with the bath water. Allowing for more refined forms of opposition, separating them, clarifying their difference, symbolizing them clearly, and understanding their consequences is exactly what a logician would and should do; this is what Aristotle did when he clarified the difference between the opposition between two contradictory statements such as A and I categorical propositions, and on the other side the weaker opposition between A and E, i.e., between contrary propositions. The clarification of such distinction is as important and appropriate to logical reasoning as is the notion of negation specific to logical systems alternative to classical logic. As Aristotle's example clearly proves, classical logicians have engaged in this type of work from the very beginning and they continue to do so.

Plumwood's final proposal is plausible and inviting: there are different logical systems and more attention should be given to the variety of systems of reasoning principles, especially if we want to have logical languages that can express more nuanced forms of negation than the negation expressed by the tilde of propositional logic. It would be hard to disagree with one such conclusion; it is a rare philosopher or logician who denies the usefulness of considering different symbolic expressions or alternative operators. When philosophers deny that a certain logical system should replace another, such as is the case with Quine's rejection of modal logic (also mentioned by Plumwood), it is usually with regard to the solution of a specific type of problems and not for wholehearted condemnation of any alternative symbolization. In the example just mentioned, Quine believed that the introduction of modal logic in quantified contexts would create more trouble than it would help resolve; hence, it was a qualified criticism within a very specific domain and should not be taken as an example of a whole rejection of any alternative system to classical logic as it is sometimes presented.[13]

11.4 Is Frege's Conception of Negation a Classical Notion?

In the previous section, I provided a preliminary critique of Plumwood's criticism of classical logic based on the notion of negation; in this section I propose to look at what Gottlob Frege, i.e., an actual "canonical" logician, said about negation and its

[13]In Plumwood's (1993) paper, no mention is made of Susan Haack's *Philosophy of Logics* (1978), whose main thesis is the existence of many logics. In her 2002 article, Plumwood presents both Haack and Quine as using the "scornful" term "deviant" for logics different from classical logic. The fact that Haack called these systems "deviant" in her pioneer book does not in my opinion indicate anything more that the contemporary labeling usage and certainly does not indicate any specific disparagement of these logics. In a web interview given on May 26, 2012, Haack states that "true rationality requires a kind of cognitive flexibility that cannot be captured in formal-logical models of scientific reasoning.".

role. My purpose is to verify whether or not this examination supports Plumwood's characterization and critique of the notion of negation as otherness.

Before examining some of Frege's statements on the notion of negation, I would like to point out the approach to logical reasoning that is apparent in the following quote from Frege (1979: 269):

> [L]anguage is a human creation; and so man had, it would appear, the capacity to shape it in conformity with the logical disposition alive in him. Certainly the logical disposition of man was at work in the formation of language but equally alongside this many other dispositions—such as the poetic disposition. And so language is not constructed from a logical blueprint.

Contrary to how Frege or classical logicians are often portrayed, this passage displays Frege's appreciation for ordinary language and his acknowledgment of its varied functions, including that of being the instrument for poetic expression. Moreover, it is clear that in Frege's view the logical ability is one of the many dispositions that humans display, which affect, but *not exclusively*, the creation of language. In this section, I aim to show that several comments of Frege's where he discusses negation indicate an understanding of it that does not support Plumwood's idea that what is negated is in any way "other" from, or dependent on, what is affirmed. These remarks are from a short essay entitled "Negation," that is part of the three articles usually collected under the title *Logical Investigations*.[14]

Frege introduces questions and negations together as variations of the way in which we may capture thoughts.[15] At this point in time, when he had made explicit the difference between sense and reference, specifically in the paper so entitled that was published in 1892, Frege regarded a thought as the sense of a sentence; famously he also thought that the truth-value of a thought was the reference of a sentence, but we do not need to concern ourselves with this feature of Frege's view that has raised several criticisms. What mattered to Frege in this essay was to reject the idea that the negation of a thought dissolves the thought in any way. He wanted to assure the existence of thoughts as separate independent entities from their expression whether in the assertion or the negation of them. Negation is parallel to assertion: they are actions that require the existence of something to be denied or asserted and this something is the thought. Thus the first important feature of Frege's notion of negation that does not fit the image of "Otherness" is the parallelism that Frege (1977: 37) repeatedly draws between assertion and negation:

[14]The first of these three articles is "Thoughts," which is the most famous of them; the second is "Negation;" these were published in 1918. The last one, "Compound Thoughts," was published in 1923. The journal was *Beiträge zur Philosophie des deutschen Idealismus*. The citations in this paper are from the reprint in Peter Geach's edition in 1977.

[15]Frege's views on thoughts are controversial for any philosophers who are not realists with regard to abstract entities. There are well known epistemological problems for such a type of realism. I am not either assuming or attacking this form of realism in this paper; I merely have to assume that for Frege's thoughts have some type of being as he explicitly states in "Negation" in order to make my points concerning his notion of negation.

> So a false thought is not a thought that has no being [...] A false thought must be admitted, not indeed as true, but as sometimes indispensable: first, as the sense of an interrogative sentence; secondly, as part of a hypothetical thought-complex; thirdly, in negation. It must be possible to negate a false thought, and for this I need the thought; I cannot negate what is not here.

False thoughts have being just like true ones and because of Frege's views on the reference of sentences, they will all have the same reference, i.e., the false. But negation is indispensable just like other words or symbols are to represent the true and the false. Frege displays the common attitude of a logician to the negation sign. Feminist critics of negation and of the logical notion of negation bestow a much greater meaning on the symbol of negation than logicians do; for the same content from a logical point of view can be expressed with and without negation. Nor is it true that logical negation brings with it any disparaging content for negated statement: "With the belief that negation has a dissolving or separating power there hangs together the view that a negative thought is less useful than an affirmative one" (1977: 40). Frege does not share this belief and uses the example of the contrapositive relation or transposition on a conditional to show this irrelevance of the negation sign. For example, "if A then B" has the same logical content than the expression "if not B then not A," as it is easily shown by replacing "A" and "B" with terms such as "mammal" and "animal" or "dog" and "animal" or by comparing "if this expense is reimbursable then it is covered by the policy" with "if this expense is not covered by the policy then it is not reimbursable." When we use either the first statement or the second in a stretch of reasoning, we can reach the same conclusion because the pairs of statements are logically equivalent; thus, Frege (1977: 41) argues,

> there is not the least ground in the nature of the case for our distinguishing between negative and affirmative premises when we are expressing the law of inference here involved. People speak of affirmative and negative judgements: even Kant does so. Translated into my terminology, this would be a distinction between affirmative and negative thoughts. For logic at any rate such a distinction is wholly unnecessary; its ground must be sought outside logic. I know of no logical principle whose verbal expression makes it necessary, or even preferable to use these terms. In any science in which it is a question of conformity to laws, the thing that we must always ask is: what technical expressions are necessary or at least useful, in order to give precise expression to the laws of this science? What does not stand this test cometh of evil.

In this clear statement of the need to separate the content from the form, Frege acknowledges that considerations regarding the emotive or symbolic meaning of negations and assertions may be appropriate in some other contexts; these distinctions may be necessary and may find ground outside of logic. But in a science—and logic is a science for Frege—only what is useful to the development of the science is necessary. Does the notion of negation as otherness usefully apply in distinguishing between "if this expense is reimbursable then it is covered" and "if this expense is not covered then it is not reimbursable?" Is the fact that the second expression contains two negated statements a reason to indict it when we can easily imagine everyday contexts in which it can usefully express a necessary condition? It is hard not to think that the interpretation of negation as otherness may have been based on examples quite different from these. However, the negation of classical logic is meant to be used in a great generality of contexts and we should not base a judgment about its alleged discriminatory force on a restricted range of uses.

Finally, a second important point about negation that in my estimation throws Plumwood's understanding of classical logic negation into doubt is the deep distrust that Frege often displayed toward the ordinary language rendition of negative statements. As Frege correctly points out, there is much ambiguity and uncertainty about how natural language expresses negation: "[I]t is by no means easy to state what is a negative judgement (thought). Consider the sentences 'Christ is immortal,' 'Christ lives for ever,' 'Christ is not immortal,' 'Christ is mortal,' 'Christ does not live for ever.' Now which of the thoughts we have here is affirmative, which negative?" (Frege 1977: 41). It is important to understand that Frege here is not merely pointing out the possible ambiguity or vagueness of ordinary language. Frege is talking about the "thoughts" expressed by these sentences, so he is pointing out how difficult it is to identify a so called "negative" thought as the unique content of information that a sentence expresses. We may be inclined to say that the sentences "Christ is mortal" and "Christ does not live for ever" express the same thought but if we accept this position, then the negation contained in the second sentence, which allegedly expresses the same thought as the first, cannot express any otherness because no such otherness is formally present or easily identifiable in the first sentence. Here, Frege does not appear to be the defender of a rigid and unyielding notion of negation, but rather a philosopher and a logician who is fully aware of the need to recognize the multifaceted nature of the human notion of negation.

Plumwood's criticism of the notion of negation in classical logic cannot be taken as representative of "the" feminist view on classical logic;[16] there is no such uniform view and there is not enough work on the topic to build the support for a homogeneous view on the subject. Frege, although considered by many the originator of modern logic, cannot be taken as the steward of a unitary received view on negation in logic. Yet, if the general claim is made that logic is somehow antithetical to women and the specific claim is made that the notion of negation is what makes logic antithetical to women, especially because of its embodying some notion of "otherness," then this section shows that at least one canonical logician does not have such a notion of negation. Whether there are other bases for the above specific and general claims remains to be seen. My ultimate aim was to stimulate the beginning of a long overdue and highly promising dialogue between the fields of logics and feminist studies.

---

[16]In her (2002) Plumwood further develops and partially alters her criticism of negation in classical logic.

11.5 A Second Major Target of Feminist Critiques: The Adversarial Method

Another canonical work in feminist literature on reason and rationality is Janice Moulton's article "A Paradigm of Philosophy: The Adversary Method." Before attempting to characterize the adversary method in positive terms, Moulton clarifies what this method should not be taken to be. Moulton does not equate competition and aggression: in the first part of her article she clearly states that aggression is neither necessary nor sufficient to participate and succeed in a competition for aggression is neither necessary nor sufficient for some features taken to be characteristics of successful participants in a competition, i.e., energy, efficiency, and competence. Neither does Moulton equate the Adversary Method with the Socratic Method; she proposes an interpretation of Socrates's strategies that justifies the ideal of Socrates as a philosopher more interested in teaching than in winning an argument. The following is a passage from a section entitled "Philosophy Reasoning—The Adversary Paradigm," in which Moulton (1983: 152–153) argues for a position similar to Nye's concerning logic and argumentation:

> I am going to criticize a paradigm or part of a paradigm in philosophy. It is the view that applies the now-rejected view of value-free reasoning in science to reasoning in philosophy. On this view all philosophical reasoning is, or ought to be, deductive. General claims are made and the job of philosophic research is to find counterexamples to the claims. And most important the philosophic enterprise is seen as an unimpassioned debate between *adversaries* who try to defend their own views against counterexamples and produce counterexamples to opposing views. The reasoning used to discover the claims, and the way the claims relate to other beliefs and systems of ideas are not considered relevant to philosophic reasoning if they are not deductive. I will call this the adversary paradigm.

I quoted this passage extensively because it displays several crucial features of some feminist approaches to logic, rationality or reason, and philosophical argumentation as it is mostly practiced in the analytic tradition.

A first important point is the allusion to the objectivity or abstractedness of the philosophical point of view. Philosophers and certainly as well logicians according to Nye and Moulton are guilty of approaching debates with a supposedly universal point of view and universal, impartial and value-free points of view are often dominant and hence "malestream" points of view. It is when we become aware of the value leadenness of any human point of view, when we accept the positionality of each of us that we can strive for a genuine objectivity borne out of the inclusion of all or at least many points of view. The extensive literature in feminist criticisms of traditional philosophy of science (see for example Harding 1986, 1991; Longino 1989, 1990, 2005; Code 1991) has produced among the most detailed and well-argued defenses of this critique.

A second point to be noticed in the above quote is the repeated identification of philosophical argumentation with the use of deductive logic. This seems to me an unusual and unsupported identification. Philosophers strive to use deductively valid and sound arguments in their writing, i.e., it is the goal of any philosopher to be able to state an argument with true premises and a conclusion that necessarily follows from the truth of those premises, but this is a lofty ideal. The sound arguments that are easy to find are not very interesting or lead to rather uncontroversial conclusions; rarely, a significant conclusion can be reached merely by using a deductive argument as can be easily verified by a survey of the great majority of philosophical writings. So, Moulton may be overstating the reach of deductive logic in philosophical argumentation. The link between the two is however very interesting from the point of view of trying to trace back the antagonism that many feminists seem to spot between reason and rationality and feminist philosophies.

A third point worth stating is the characterization of the paradigm of a philosophical discussion as *adversarial*. There is truth in this given that in philosophy usually the debate centers around a thesis being advanced or argued for on the basis of premises; the person who wants to deny a thesis is encouraged to attack either the inferential form of the argument, i.e., whether or not the premises provide reasons to accept the truth of the conclusion, or the accuracy of the claims made by the premises. This is a dialogical pattern of assertions and denials, in a back and forth debate. Yet, the pattern does not "require" nastiness or aggression just like competition does not require aggression, as Moulton acknowledges. Indeed, there have been many philosophers who have written on this paradigm asserting how the language often used in describing the adversarial method unduly portrays a war-like picture of philosophical argumentation that is not the most accurate depiction of what actually happens in a philosophical discussion (Cohen 2013). Most likely, a philosophical debate needs to be dialogical or dialectical, with theses asserted and denied and with examples and counterexamples used to support the theses and their denials, but such debate need not be *adversarial*. As Daniel Cohen aptly points out, discovering a mistake in one's reasoning or belief is a gain or a victory rather than a loss.

Finally and naturally ensuing from the previous point is the interesting attack on the method of counterexamples that Moulton identifies as central to philosophical debate. At first glance it may not seem clear why Moulton connects counterexamples with argumentation and with deductive reasoning, so let us look at an example to get clear on what she may have meant. Suppose I want to argue for the claim that (a) all students who passed my logic course with an A grade are females. This is a universal generalization: it starts with "all" which is the word we use when we want to talk about a whole class of objects, in this example, all students who passed my logic course with an A grade. "All" is very different from "most" of course since this latter may be true even if not all five female students in my logic course passed the course with an A grade but if only three or four of them passed it earning an A grade. What would make this statement false? Clearly if Alex, who is also a student taking my logic course and who identifies as a male student, passed the course with an A grade. Alex's success is the counterexample to the truth of (a). Notice that we can easily extrapolate a rule from this example: all statements of the form (a), all A are B, is shown to be false if we find an A which is not B. We may falsify any generalization like (a), by finding a counterexample.

As it should by now be pretty clear, the stretch of reasoning just illustrated is fairly common and usually tacit; our brain and our consciousness do not spend, or waste, much time wondering about the truth of generalizations such as the one I gave above. On the other side, the generalization "all US policemen tend to profile a male black man as likely to be dangerous" is a generalization that would greatly worry us if we had evidence supporting its truth. We would not want one such alleged truth to go unchallenged and we would look for counterexamples to support its denial. Perhaps the first generalization may seem more innocuous or lacking in concerning consequences; and yet, it too might worry my dean if it were used to prove that I do not give the same chance of succeeding in my logic course to both female and male students.

I find all the points Moulton raises in this passage interesting and deserving to be discussed. Where I disagree with Moulton is when she seems to identify the Adversary Method with deductive reasoning and when she describes this latter as the main form of reasoning that is employed in the scientific fields. The second part of this conjunction is inaccurate; the first is puzzling and probably rests on a broader notion of "deductive reasoning" than I employ.

Moulton brought attention to some important features of philosophical argumentation and practice that have been considered possible causes of the low number of women engaged in philosophy. Here, I do not develop a more thorough discussion of the merit of Moulton's claims; instead I conclude by looking at a recent discussion of the adversarial method that takes its starting point from Moulton's critique. First of all, Rooney (2012: 321) correctly points out that "few meaningful non trivial arguments in philosophy (or elsewhere) are valid and sound;" after mentioning the fact that discomfort with the adversarial style of argumentation is often mentioned as a factor at the origin of many women's avoidance of philosophical debates, she also states:

> [F]ramed largely in terms of gender differences, discussions about adversarial argumentation regularly devolve into all-too-familiar debates about gender essentialism, and about the possibility of offending women who are quite comfortable and effective with an adversarial style of argumentation—or might be encouraged to be so. My main concern is with the way in which these debates can function as red herrings, deflecting attention from other critical examinations of adversarial argumentation in philosophy, including, especially, other feminism-informed ones that draw attention to diversity as a concern that is not just confined to gender inequity (Rooney 2012: 317).

Two main concerns of the previous attack on the adversarial method emerge in Rooney's discussion: (1) gender essentialism, and (2) the neglect of other aspects of the adversarial approach to argumentation that negatively affect other subjects that cannot be identified by merely focusing on the category of gender.

Essentialism in philosophy is not a popular position, often the target of crushing criticisms. This is true also within feminist discussion; the view that all or most women share certain characteristics because of their shared biology is not supported by many feminists and it can dangerously degenerate into the view that biological features determine the social positions of women and men. Longino (1990: 185) cites Bleier (1984), Rose (1983), and Harding's earlier work (1980) as examples of views that claim that "women have certain traits (for example, dispositions to attend to particulars and interactive and cooperative social attitudes and behaviors rather than individualistic and controlling ones) that enable them to understand the true character of natural processes (which are complex and interactive)." But Longino (1990: 188) points out, "the characterization of feminist science as the expression of a distinctive female cognitive temperament has other drawbacks, the greatest being that it conflates feminine with feminist. While it is important to reject the traditional derogation of the virtues assigned to women, it is also important to remember that women are *constructed* to occupy positions of social subordinates. We should not uncritically embrace the feminine." Longino's is an apt reminder with regard to rationality, reason and logical argumentation, too. Once the social construction of gender and sex is accepted, essentialism becomes less and less appealing while trying to understand why femaleness and femininity have been systematically opposed to rationality and reason. As Rooney points out, essentialism is also a view that is falsified by the many women who do enjoy dialectical argumentations[17] and that has possibly the negative effect of discouraging women who enjoy a good philosophical debate from entering into the discussion for fear of being regarded as unfeminine.

Yet, the task of explaining how mathematics or logic could be socially constructed is not an easy one: in these disciplines, the study of cultural differences has not provided much evidence of significant differences in the ways in which humans reason. It is exactly the apparent necessity of mathematical and logical truths and their presumed self-evidence to all those who have fully understood their meanings that discourage a social constructivist explanation of their origin. Wittgenstein's philosophical views of mathematical truth as arising from well-established practices and rules has been one of the very few views that envisions a grounding of mathematics in a non-Platonist and non-realist origin. How this very view has been criticized and dismissed may perhaps give an idea of how challenging it would be for a feminist to argue for the social origin of mathematical and logical truths. However, it seems equally unlikely that a feminist could embrace Platonism, i.e., the view that the truths of logic and mathematics reflect the reality of abstract objects. It would seem somewhat incoherent to claim that logic, negation, identity, etc. are affected by the categories of social power and discrimination and yet accept that they reflect some mind independent entities. It would seem more plausible for feminists to claim that the features of logic that separate and exclude women are the result of human patriarchal construction, rather than an inescapable feature of some mind-independent reality.[18]

[17]If we consider the fact that the majority of law degrees in the US are granted to women and that the study and practice of the law requires extensive use of argumentation, the myth that women do not study analytic philosophy because of its adversarial method of argumentation flounders miserably.

[18]Recent work in experimental philosophy (see Buckwalter and Stich 2014) supports the claim that philosophical intuitions may be socially and culturally constructed.

The second point raised by Rooney reflects the long standing effort of feminist philosophers to broaden their critical discussion by taking into account the intersectionality of systems of oppression. Essentialism, by focusing only on the gender difference as the origin of women's alleged discomfort with logical reasoning and philosophical argumentation, ends up neglecting the possibility that there may be other causes affecting who enjoys and who does not enjoy a rational and dialectical approach to a discussion. The second part of Rooney's discussion intriguingly develops an examination of how a skeptical stance may be more or less nurturing and helpful to a minority opinion. This is an example of a valuable, original, and thoroughly feminist friendly approach to the question of how can women and other socially discriminated groups claim back reason and rationality and enjoy the practice of philosophy.

Even from the two very limited examples I outlined, i.e., the notion of negation and the adversarial method of philosophical argumentation, it should be clear that feminist critics have brought to light rarely discussed aspects of philosophical reasoning and argumentation. Reflecting critically on the work of pioneers like Nye, Plumwood or Moulton is the only way to take it seriously and build on their quite valuable original efforts. The interaction between feminist criticisms of traditional notions of reason or rationality and the rereading of classical works on reason or rationality promises to produce broader and stronger notions.

Acknowledgments I thank Kari Hanson and Tessa Hagen for their useful comments on both content and bibliography.

References

Agonito, R. (1977). *History of ideas on women*. New York: Berkley Publishing Group.

Antony, L. (2012). Different voices or perfect storm: Why are there so few women in philosophy? *The Journal of Social Philosophy, 43*(3), 227–255.

Ayim, M. (1995). Passing through the needle's eye: Can a feminist teach logic? *Argumentation, 9*(5), 801–820.

Bleier, R. (1984). *Science and gender: A critique of biology and its theories on women*. New York: Pergamon.

Buckwalter, W., & Stich, S. (2014). Gender and philosophical intuitions. In J. Knobe & S. Nichols (Eds.), *Experimental philosophy* (Vol. 2, pp. 307–345). Oxford: Oxford University Press.

Butler, J. (1990). *Gender trouble*. London: Routledge.

Code, L. (1991). *What can she know?*. New York: Cornell University Press.

Cohen, D. H. (2013). Daniel H. Cohen: For argument's sake. TED video, 9:35, from presentation February 2013, posted by TEDxColbyCollege, February 2013. https://www.ted.com/talks/daniel_h_cohen_for_argument_s_sake?language=en/. November 6, 2014.

De Beauvoir, S. (1949). *Le deuxième sexe*. Paris: Gallimard.

Dennett, D. (1976). Conditions of Personhood. In A. Rorty (Ed.), The Identity of Persons (pp. 175–196). Berkeley: University of California Press.

Fausto-Sterling, A. (1985). *Myths of gender: Biological theories about women and men*. New York: Basic.

Frege, G. (1977). Negation (Or. Pub. 1918). In P. Geach (Ed.), *Logical investigations* (pp. 31–53). New Haven: Yale University Press.

Frege, G. (1979). Sources of knowledge of mathematics and the mathematical natural sciences (Or. Pub. 1924/25). In P. Long, R. White (Trans.) *Posthumous writings* (pp. 267–274). Oxford: Basil Blackwell.

Gatens, M. (1991). Feminism and philosophy: Perspectives on difference and equality. Bloomington: Indiana University Press.

Green, K. (1995). *The woman of reason: Feminism, humanism and political thought*. Cambridge: Polity Press.

Haack, S. (1978). *Philosophy of logics*. Cambridge: Cambridge University Press.

Haack, S. (2012). Susan Haack on the current state of professional philosophy, interview by R. Carrier. http://www.patheos.com/blogs/camelswithhammers/2012/05/susan-haack-on-the-current-state-of-professional-philosophy/#ixzz3F6FVSuai/. Accessed November 6, 2014.

Harding, S. (1980). The norms of inquiry and masculine experience. In P. Asquith & R. Giere (Eds.), *PSA 1980* (pp. 305–324). East Lansing, MI: Philosophy of Science Association.

Harding, S. (1986). *The science question in feminism*. Ithaca, NY: Cornell University Press.

Harding, S. (1991). *Whose science? Whose knowledge? Thinking from women's lives*. Ithaca, NY: Cornell University Press.

Haslanger, S. (2008). Changing the ideology and culture of philosophy: Not by reason alone. *Hypatia, 23*(2), 210–223.

Hass, M. (2002). Fluid thinking: Irigaray's critique of formal logic. In R. J. Falmagne & M. Hass (Eds.), *Representing reason: Feminist theory and formal logic* (pp. 71–88). Lanham: Rowman and Littlefield.

Heikes, D. K. (2012). *The Virtue of Feminist Rationality*. London: Bloomsbury.

Heikes, D. K. (2010). *Rationality and feminist philosophy*. London: Continuum.

Hubbard, R. (1990). *The politics of women's biology*. New Brunswick, NJ: Rutgers University Press.

Hutchinson, K., & Jenkins, F. (2013). *Women in philosophy: What needs to change?*. New York: Oxford University Press.

Lloyd, G. (1984). *The man of reason "male" and "female" in western philosophy*. London: Routledge.

Locke, J. (1975). *An essay concerning human understanding*. Oxford: Clarendon Press.

Longino, H. (1989). Feminist critique of rationality. *Women's Studies International Forum, 12*(3), 261–269.

Longino, H. (1990). *Science as social knowledge*. Princeton: Princeton University Press.

Longino, H. (2005). Circles of reason: Some feminist reflections on reason and rationality. *Episteme, 2*(1), 79–88.

Martin, E. (1991). The eggs and the sperm: How science has constructed a romance based on stereotypical male-female roles. *Signs, 16*(3), 485–501.

Morgan, F. (1989). *A misoginist's source book*. London: Jonathon Cape.

Moulton, J. (1983). A paradigm of philosophy: The adversary method. Reprint in S. Harding, & M. Hintikka (Eds.), *Discovering reality* (pp. 149–164). Dordrecht: Springer.

Noonan, H. (2003). *Personal identity* (2nd ed.). London: Routledge.

Nye, A. (1990). *Words of power: A feminist reading of the history of logic*. London: Routledge.

Paxton, M., Figdor, C., & Tiberius, V. (2012). Quantifying the gender gap: An empirical study of the underrepresentation of women in philosophy. *Hypatia, 27*(4), 949–957.

Plumwood, V. (1993). The politics of reason: Toward a feminist logic. *Australasian Journal of Philosophy, 71*(4), 436–462.

Plumwood, V. (2002). Feminism and the logic of alterity. In R. J. Falmagne & M. Hass (Eds.), *Representing reason: Feminist theory and formal logic* (pp. 45–70). Lanham: Rowman and Littlefield.

Prokhovnik, R. (1999). *Rational woman: A feminist critique of dichotomy*. London: Routledge.

Quinton, A. (1962). The soul. *Journal of Philosophy, 59*, 393–403.

Rooney, P. (2012). When philosophical argumentation impedes social and political progress. *Journal of Social Philosophy, 43*(3), 317–333.

Rose, H. (1983). Hand, brain, and heart: A feminist epistemology for the natural sciences. *Signs, 9*(1), 73–90.
Shoemaker, S. (1970). Persons and their pasts. *American Philosophical Quarterly, 7*, 269–285.
Tyson, S. (2014). From the exclusion of women to the transformation of philosophy: Reclamation and its possibilities. *Metaphilosophy, 45*(1), 1–19.
Van Camp, J. (2006). Deep thought: For (mostly) men only? Does it matter? Manuscript.
Warren, K. (1990). The power and promise of ecological feminism. *Environmental Ethics, 12*, 125–146.

Chapter 12
Feminist Versus General Philosophy of Science

Raffaella Campaner and Maria Carla Galavotti

In the last few decades the literature on feminism and philosophy of science has flourished and continues to raise a vast and varied array of issues, calling attention to a number of ways in which dominant science is gender biased. Philosophy of science, especially its "received view," has been deemed similarly prejudiced. However, in the last half-century philosophy of science has undergone substantial changes, and is now better equipped to address the issues raised by feminist epistemologists. On the one hand, feminist philosophy has moved from the intent of fighting underrepresentation and eradicating sexist bias in research and scientific activities, to a more general reconsideration of modes of knowledge. On the other, recent philosophy of science has been largely guided by the aim of faithfully representing the production of scientific knowledge as it is actually performed in the sciences, with an eye to disciplinary and contextual specificities. On that basis, there seems to be vast ground for cooperation between the two. In particular, Helen Longino's contextual empiricism offers some suggestions that can both converge on and reinforce some current trends in philosophy of science.[1]

12.1 Feminist Epistemology and the Bottom-up Approach to the Philosophy of Science

In the chapter "Feminist Epistemology and Philosophy of Science" of the Stanford Encyclopedia of Philosophy, Anderson (2011) gives an extensive survey of the literature on feminism and science. She summarizes the charges moved by feminists to official science as follows:

[1]Sections 12.1 and 12.2.3 are by Maria Carla Galavotti; Sects. 12.2.1, 12.2.2 and 12.3 are by Raffaella Campaner.

R. Campaner (✉) · M.C. Galavotti
University of Bologna, Bologna, Italy
e-mail: raffaella.campaner@unibo.it

M.C. Galavotti
e-mail: mariacarla.galavotti@unibo.it

M.C. Amoretti and N. Vassallo (eds.), *Meta-Philosophical Reflection on Feminist Philosophies of Science*, Boston Studies in the Philosophy and History of Science 317, DOI 10.1007/978-3-319-26348-9_12

> Various practitioners of feminist epistemology and philosophy of science argue that dominant knowledge practices disadvantage women by (1) excluding them from inquiry, (2) denying them epistemic authority, (3) denigrating their "feminine" cognitive styles and modes of knowledge, (4) producing theories of women that represent them as inferior, deviant, or significant only in the ways they serve male interests, (5) producing theories of social phenomena that render women's activities and interests, or gendered power relations, invisible, and (6) producing knowledge (science and technology) that is not useful for people in subordinate positions, or that reinforces gender and other social hierarchies. Feminist epistemologists trace these failures to flawed conceptions of knowledge, knowers, objectivity, and scientific methodology.

Feminist epistemologists react to these biases by proposing new conceptions of knowledge, scientific methodology, theory formation and evaluation. The general idea is to replace the conception of knowledge as "pure" with *situated knowledge*, namely knowledge located within the perspective of knowledge producers and users. Various tendencies have emerged within feminist epistemology. In spite of relevant divergences among them, a number of tenets appear to be largely shared—albeit to a varying extent and emphasis. These include some kind of *pluralism* rooted in the under-determination of scientific theories by evidence, which goes hand in hand with a rejection of the idea that scientific knowledge can qualify as *strongly objective*, or objective in a way that transcends the knowing subjects. The strict link between *cognitive* and *practical* aspects of science is stressed with a rebuttal of a sharp separation between a context of discovery and a context of justification. In addition, feminist philosophers of science emphasize the *contextual* character of scientific research and its dependence on all sorts of assumptions and social constraints, resulting in a dynamic view of knowledge.

While these aspects are largely overlooked within the "received view" of the philosophy of science shaped by logical empiricism, the same cannot be said for present day philosophy of science, which results from the deep changes that started in the Sixties and are still underway. It is well known that the so-called *post-positivist* turn, usually associated with authors like Thomas Kuhn, Imre Lakatos, Paul Feyerabend, Norwood Russel Hanson and many others, opened philosophy of science to the historical, psychological and sociological elements connected with the formation and growth of scientific knowledge. Although welcoming such an opening, feminist epistemologists urge the need to go further by addressing the gender issue and embracing a new concept of objectivity.

In parallel with the post-positivist development, a no less important turn has been brought about by Patrick Suppes, Bas van Fraassen, Frederick Suppe, Ronald Giere, and many others, who replaced the syntactical approach to scientific knowledge embraced by logical empiricists with a more flexible pluralistic attitude focusing on models rather than theories and emphasizing the contextual character of science.[2] A key role in such a turn is to be ascribed to Suppes' pioneering work, which is by and large ignored by the literature on feminism, surprisingly so given

[2]See Suppe (2000) for an account of the shift from the received view of theories to the so-called semantic approach.

the striking affinity between his conception of philosophy of science and some of the claims emerging in the feminist literature. Suppes' book entitled *Probabilistic Metaphysics*, published in 1984 but still very fresh, does not seem to have received the attention it deserves not only on the part of feminists, but also by philosophers of science at large. The book is a vehement attack on what he calls the chimeras of the traditional view of rationality, namely completeness of knowledge, certainty and determinism. Against the ideal of completeness, namely the conviction that "scientific knowledge can in principle be made complete" (Suppes 1984: 2), Suppes holds that "the collection of past, present, and future scientific theories is not converging to some bounded fixed result that will in the limit give us complete knowledge of the universe" (Suppes 1984: 10). The ideal of certainty of science is also discarded on the grounds that "certainty of knowledge—either in the sense of psychological immediacy, in the sense of logical truth, or in the sense of complete precision of measurements—is unachievable" (Suppes 1984: 10). Both certainty and the related deterministic view that the future is determined by the past, and that every event follows from some cause, should leave room for the awareness that scientific knowledge is pervaded with uncertainty. Notably, Suppes is led to such claims by looking at science from an angle different from that taken by logical empiricists, who located the context of discovery out of the realm of philosophy of science. By contrast, Suppes believes that the distinction between context of discovery and context of justification should be abandoned in favor of an integrated view of scientific knowledge that does not trace a sharp separation between theory and practice. Suppes' attitude in this connection, which is nowadays largely accepted by philosophers of science, is a recurrent theme of the feminist literature.

Another crucial aspect of Suppes' approach is the view that scientific knowledge has an irreducibly *local character* and is to be analyzed *within a specific context*. In Suppes' words: "like our own lives and endeavours scientific theories are local and are designed to meet a given set of problems" (Suppes 1981: 14–15). The idea here is that scientific knowledge is not amenable to a unique characterization because the complexity of phenomena and the variety of practical situations in which phenomena are investigated are such that important notions in science—and in philosophy as well—cannot be forced into some definition given once and for all. Awareness of this cannot but suggest a *pluralistic attitude* that admits of different ways of representing phenomena and the adoption of different methods of investigation. Rather than being a form of relativism, this kind of pluralism urges the need to take into account the context in which scientific knowledge is formed, and to make explicit its underlying assumptions. As acknowledged by Kellert, Longino and Waters in the introduction to the collection *Scientific Pluralism*(see Kellert, Longino and Waters (Eds.) 2006), the pluralistic attitude often stressed by the feminist literature was anticipated by Suppes, who titled his presidential address to the Philosophy of Science Association delivered in 1978 "The Plurality of Science" (see Suppes 1981).

Suppes' approach embodies a shift of attention from laws to models. At the turn of the Sixties he put forward a proposal to overcome the traditional view of theories upheld by logical empiricists, based on the distinction between theoretical and observational language, in favor of a model-centred conception according to which the relation between empirical theories and data "calls for a hierarchy of models" (Suppes 1962: 253).[3] A crucial aspect of this viewpoint is the conviction that theoretical and observational models are not to be conceived separately, but engaged in a continuous interplay. This led Suppes to adopt a strongly anti-reductionist standpoint with regard to all scientific fields, from mathematics to physics and the social sciences, in which connection he holds that "any program for the elimination of unobservable theoretical variables would seem […] to be completely mistaken" (Suppes 1990: 27). A further issue emphasized over and over again in Suppes' writings is the role of assumptions in the construction of scientific knowledge. Indeed, models result from a combination of empirical evidence with assumptions of various nature, so the adoption of a model-centered approach involves recognition of the role played by the assumptions that are made at all levels of model building, and urges the need to make them explicit.

Anti-reductionism is a leitmotif of the literature on feminist epistemology, strongly stressed among others by Longino (1990: Chap. 10) who calls attention to the link between pluralism and anti-reductionism.[4] By emphasizing the social dimensions of scientific knowledge, Longino's *social constructionism* implies both a commitment to pluralism and a rejection of reductionism. On the one hand, she argues for a theoretical pluralism that "accepts a variety of theories on a given subject matter. Which theory we use to guide our interactions with or interventions in natural processes, which we appeal to in providing explanations depends on our interests at a particular time" (Longino 1990: 230–231). On the other hand, she claims that the view of scientific knowledge as social knowledge is anti-reductionist in a twofold sense. First, "scientific knowledge cannot be reduced to the knowledge of an individual […] Secondly, an individual's scientific knowledge is made possible by that individual's social and cultural context" (Longino 1990: 231). The emphasis placed by Longino on the social and cultural context in which scientific hypotheses are formed is a major achievement of feminist epistemology. A number of researchers have in fact argued that in several fields male-biased assumptions have played a decisive role in shaping explanations and theories, leading in some cases to wrong results. Take for instance Elizabeth Lloyd's analysis of the explanation of female sexuality in relation to reproduction. The strict link between female orgasm and reproduction established by a widespread view is shown by Lloyd to be based on the male-biased assumption that "female sexuality functions completely in the service of reproduction" (Lloyd 1996: 92), an assumption that was long retained in spite of evidence to the contrary. In full agreement with Longino, Lloyd maintains that "social assumptions and prior commitments of the scientists play a major role in the practice of science itself, at many levels—experimental design, data collection, predictions, hypotheses formulation, and the evaluation of explanations" (Lloyd 1996: 100).

---

[3]The most extensive treatment of Suppes' view of theories and of his philosophy of science at large is to be found in Suppes (2002).

[4]It is noteworthy that in connection with anti-reductionism Longino mentions Suppes.

A model-centred approach to the philosophy of science which takes models to provide the proper tool for describing, explaining and predicting phenomena is today widely adopted by philosophers of science. In this spirit, and taking inspiration from Suppes' work, a number of writings by one of the authors of the present chapter heralded a *bottom-up approach* which seems apt to accommodate at least some of the issues raised by the feminist literature (see especially Galavotti 2006, 2014). That a model-based perspective has significant implications for feminism has been observed by Angela Potochnik, who embraces a thesis that we find congenial, namely that "model-based science and socially engaged science go hand in hand nicely" (Potochnik 2012: 388). Oddly enough, her paper "Feminist Implications of Model-based Science" mentions a number of authors, including among others Frederick Suppe, Ronald Giere, Bas van Fraassen, Peter Godfrey-Smith, Nancy Cartwright, and James Woodward, but does not mention Suppes at all.

Let us now see how the notions of explanation, prediction and causality are conceived within the bottom-up approach we are defending. In the Seventies, one of the first to embrace a pluralistic, model-based view of explanation, Bas van Fraassen put forward a *pragmatics of explanation*, which regards explanation as a "three-term relation, between theory, fact, and context" (van Fraassen 1980: 156). Accordingly, explanation is an answer to a why-question posed by somebody to somebody else, and what kind of answer is sought and offered depends on the information available to the agents involved, the purpose for which explanation is requested, and the use the questioner intends to make of it. A corollary of van Fraassen's pragmatical view, which epitomizes the concept of explanation retained by the bottom-up approach, is that the conceptual setting surrounding the phenomena under study is part of the context. The flexibility of this perspective makes it naturally apt to accommodate the suggestions coming from feminism. As observed by Potochnik (2012: 388): "a feminist approach to explanation should acknowledge the context-dependence of explanations and investigate how social values shape the contexts." This claim is obviously in tune also with Longino's social constructionism.

The bottom-up approach embodies a model-based notion of causality which is genuinely context-dependent. Also in that connection Suppes has done innovative work by developing a view that does not contemplate "genuine" causes, leaving to the context the specification of what is to be taken as causal; in a pluralistic spirit he regards any attempt at working out a notion of causality applicable to every situation in which causal attributions are made as deemed to failure because different contexts call for a different characterization of causality (see Suppes 1970). Suppes' view of causality is void of ontological content, being rather linked to invariance: causal interpretation applies to models exhibiting a certain degree of invariance, and can therefore be used to plan and perform interventions. A suchlike conception of causality, having more to do with prediction and intervention than with explanation, has a long tradition in a number of fields including political economy, epidemiology, and management.[5] Without going into further details, it is worth stressing that a crucial component of the bottom-up approach is the idea that explanation and causality are *situated within a given context*, can serve different purposes and can be performed at varying levels of generality and detail. A perspective like the one just described seems to offer fertile ground for incorporating the claims advanced by feminists. Authors like Potochnik and Longino indicate this as the way to go. The stress put by feminists on social factors and the role they can play in connection with explanation and prediction is bound to bring new insight into the debate on such topics, and more work in this direction will hopefully contribute to a better integration between the feminist literature and the perspective we have been discussing.

It is worth noting that a bottom-up approach has recently been taken also by authors working on the foundations of mathematics and statistics. A view of the foundations of mathematics endowed with a strong pragmatical flavor is heralded by the logician Carlo Cellucci, who holds that mathematics is developed from below; it pursues the solution of some problem by formulating hypotheses based on evidential data, making use of assumptions and inferential methods that call for a contextual justification. Cellucci ascribes a key role to the notion of *plausibility*, which is not a priori but based on experience: a hypothesis qualifies as plausible if "comparing the arguments for and the arguments against the hypothesis on the basis of the existing data, the arguments for must be stronger than those against" (Cellucci 2013: 34). The process of knowledge acquisition is both one of discovery and one of justification, so there is no separation between the two. The justification of methods and assumptions relies on the success of the hypotheses that have been formulated, not on their alleged truth. The kind of justification being called for is *vindication*: "both deductive and non-deductive rules can be vindicated with respect to an end that agrees with their role in knowledge" (Cellucci 2011: 137). Cellucci makes a strong point against the traditional concept of knowledge as "true justified belief," inspired by the ideals of truth, objectivity and certainty, to which he opposes a heuristic conception of knowledge aiming at *plausibility* rather than truth, and *agreement* rather than objectivity.

In a similar vein, Christopher Hennig heralds a bottom-up approach to mathematical, and more specifically statistical modeling. He regards mathematical modeling as a *social construct*, albeit one of a peculiar kind because of its capacity to deliver stable representations of phenomena which meet with universal agreement. Like Cellucci, Hennig identifies the origins of mathematics with the practice of counting aimed at the satisfaction of primary needs. In the course of its history, mathematics evolved towards increasing abstraction, up to a point where it has acquired a reality of its own independent of its origin. At that point, mathematics was endowed with a strongly objective character and associated with the ideal of absolute truth. From the point of view of Hennig's social constructivism there is no need for such a strong, realistic conception of truth in order to understand or justify

[5]See Campaner and Galavotti (2007, 2012), Galavotti (2008) for more on the concept of causality in a pluralistic perspective.

mathematical modeling: "the idea of absolute truth in mathematics can be explained by a historical process of construction that made binding agreement the essential aim of mathematical communication" (Hennig 2009: 42; see also Hennig 2010). Mathematical models serve a number of practical purposes such as prediction, manipulation, and explanation, and they reduce complexity, support decision-making, and improve mutual understanding. The justification of models is given by their success in fulfilling the tasks they have been devised and adopted for. Concentrating on statistics, Hennig calls attention to the assumptions underlying statistical methods, warning against the widespread tendency to "quantify evidence in a unified way regardless of the subject matter" (Hennig 2009: 44). Against such a practice, which he deems rooted in the conviction that the mere application of statistical methods to data uncritically taken as "given" can produce "objective" results, Henning advocates the adoption of a bottom-up approach that starts from the context in which data are collected to move on to the formation of models representing them and their use for various purposes, recommending that at each step of this process all the assumptions that are made are spelled out and justified in view of the aim of inquiry.

The standpoint taken by Cellucci and Hennig is of special interest in the context of the present discussion because of the emphasis they put on the social origin of mathematical truth and the contextual character of mathematical modeling. In that sense, their work complements the bottom-up approach to the philosophy of science outlined above. As already argued, an approach like the one described could meet at least some of the concerns arising from the feminist literature. In particular, the bottom-up approach is apt to be combined with the *contextual empiricism* developed by Helen Longino. Her position is empiricist in that it treats experience as the basis of all knowledge claims in the sciences, and it is contextual in that it stresses the relevance of the context—including assumptions supporting reasoning processes and social and cultural values—in the construction of knowledge. The aim is a deep understanding of the role of contextual factors, conceiving scientific knowledge as the result of "complex interactions of what is known, what is assumed, and *social-cognitive values*" (Longino 1990: 186, italics added). According to this social account of objectivity, scientific knowledge emerges as the product of collaborative social interactions among scientists, and retains the marks of culture and context-dependent assumptions. Longino's standpoint is well summarized by the following passage, taken from one of her less quoted articles: "That any theory is judged to have greater evidential support than alternatives may be (usually is) a function of shared assumptions in the community that have made certain data salient or worthy of attention and/or that devalue or fail to make perspicuous other data or that frame problems in particular ways. Objectivity, insofar as it makes sense to speak of it, would instead consist in the critical scrutiny of data, reasoning, and assumptions, that is of elements that go into the construal of evidential relations, by a scientific community that includes multiple perspectives and whose discursive interactions satisfy the norms of critical contextual empiricism" (Longino 2006: 173).

12.2 Feminist Epistemologies and the Sciences

Resting on the considerations elaborated above, this section shall focus on specific areas to which feminist discourse on science has been significantly applied in different ways. We shall start by highlighting some appreciable points of convergence between feminist philosophy of science and the current debate in the philosophy of the life sciences, having to do especially with modeling biological and biomedical phenomena (Sect. 12.2.1). These are clearly *gendered objects*, and can hence figure as the natural object of feminist investigation. We will then shift to chemistry, an apparently gender-immune hard science, and discuss some investigations on *historical* and *methodological* issues in modeling gases carried on from a feminist perspective (Sect. 12.2.2). Finally, we shall consider feminist criminology (Sect. 12.2.3), a field where the feminist input has opened new perspectives.

12.2.1 Feminist Philosophy of Science and Modeling in the Life Sciences

As pointed out in the first section, both feminist philosophy of science and contemporary general philosophy of science pay much attention to the elaboration of models in the sciences, and to the different roles models can play in different contexts and with respect to different cognitive aims. The representational, explanatory and predictive roles of models, their levels of abstraction and idealization, epistemic commitments and trade-offs between generality and accuracy are topics on which both feminist epistemologies and current philosophy of science have been widely resonating. As stressed by Potochnik (2012), the two fields have crucial resources to offer to each other's project in these respects, both aiming, amongst others, at de-emphasizing truth as a scientific target and stressing the perspectival nature of knowledge. That doing science is deeply grounded on the construction of models, that multiple models are used to represent and explain the same phenomena, and that modeling desiderata and assumptions must be made explicit are among the features of scientific practice that—as Potochnik points out—have been recognized in parallel both by feminist and general philosophy of science, and hence among those on which they can fruitfully dialogue.

Ontological heterogeneity of objects of investigation, complexity, and mutuality of interactions between component parts of phenomena explored have been deemed worthy of deep analysis, and contrasted with so-called "superempirical values" like, for instance, simplicity and generality, which have been more commonly pursued by mainstream accounts of scientific theories. As remarked in Sect. 12.1, Longino has advanced her contextual empiricism as an epistemological position focusing on evidential reasoning, cognitive practices, and the role of the context and background assumptions in the elaboration of knowledge. Referring specifically to biological phenomena, she points out that feminist researchers "have resisted unicausal accounts of development in favor of accounts in which quite different factors play causal roles. They, therefore, emphasize the multiplicity of kinds of factors at all developmental levels, from within the cell to the whole organism" (Longino 2008: 71). Feminist philosophy of science has played a remarkable role in endorsing the relevance of a whole spectrum of notions. It is however undeniable that these cannot be exclusively ascribed to feminist philosophy of science. A focus on complexity, reciprocal relations, and multiplicity of levels, perspectives and cognitive aims has characterized to different extents recent inquiries on modeling biological phenomena by leading philosophers of science. To mention just a few, we can recall William Wimsatt's reflections since the Seventies on the evolution of complex, functionally organized systems, on the limits of reductionist methodologies, and on the importance of interlevel accounts. Wimsatt has brought into focus heuristics and models that are employed in science practice, to be always conceived as having local goals and methods (see Wimsatt 1974, 1994, 2006).

That the biological world it too complex and messy to be successfully organized by means of a few inclusive principles has also been argued by Schaffner (1993), who suggests that just "middle-range theories" can be identified, and "piecemeal reductionism" pursued. Not only the promises, but also the limits of reductionism in the biomedical sciences have been largely addressed also, for instance, in a collection of essays edited by Van Regenmortel and Hull (2002). Within successful neo-mechanist positions such as Craver's and Darden's (2013), such notions as "interlevel relation" and "interfield account" have been widely investigated, and the crucial role of organization and interactions between levels has been stressed. The relevance of context in the choice of models has been pointed out as well: "when biologists identify mechanisms, there is *an inherent perspectival aspect* as to what is picked out of interest from all the goings on in the world. First, the choice of phenomenon is relative to the scientist's interests. [...] To some extent and in some cases, the choice of beginning, ending, topping-off, and bottoming-out points in the description of a mechanism may also be related to the interests of the investigator" (Darden 2008: 960, italics added). All these suggestions converge towards the idea, supported by Longino, that local epistemologies allow us to conceive of situations characterized by a plurality of theories and models, and to insert them into a pluralist framework which is not taken as provisional, as a sign of the immaturity of a given field of research, but, at least in principle, as one of its constitutive features.

From the vast literature dealing with modeling biological systems it clearly emerges that a few issues in model building have been recognized as crucial *both* within feminist philosophy of science *and* general philosophy of science. The ontological heterogeneity of some domains, the need to model many kinds of properties and behaviors, and the building of multicausal accounts have been largely tackled in the last few decades in fields as the philosophy of biology and the biomedical sciences, without any direct commitment to gender stances. Given their affinity on these topics, some joint effort of feminist philosophy and philosophy of the life sciences might be particularly fruitful. The life sciences provide very interesting and challenging issues in a number of respects. On the one hand—as shown, for instance, by feminist studies on genetics, endocrinology, evolutionary studies and primatology (see e.g. Keller 1983; Haraway 1986; Martin 1991; Sperling 1991; Fausto-Sterling 2000), as well as by those on female sexuality recalled above—the very *object* of some biological lines of research is inherently characterized by gender aspects, and hence their investigations have been deeply affected by gender issues. On the other hand, a few general aspects of the life sciences—such as their dealing with complex, highly variable and exception-ridden phenomena—have led to bring into focus their methodological features and to differentiate them from those of such sciences as physics[6] and chemistry.

12.2.2 Gender Issues and Chemistry

We believe it is worth stressing how gender issues can affect not only sciences directly dealing with gendered organisms, but also the so-called hard sciences. Reflections on gender can significantly enter not only into a discourse on the *object* of scientific investigation, but into a *historical* discourse on scientific *methodologies*. In the context of chemistry, whose object cannot be attributed a gender, some fairly recent works take scientific knowledge on ideal gas, and related laws, to be analyzable from a gender perspective. Responding to the challenge of showing that even something like Boyle's law can contain some gender bias, Potter (2001) investigates its origins in depth, showing that the law was strongly conditioned by a specific scientific and cultural context, and hence determined also by class and gendered considerations. Potter stresses how all scientific knowledge is culturally and historically situated, and how gender politics can turn out to be embodied also in a well-established principle, well-grounded in experimental evidence: had it been only for the experimental evidence available at the time—she maintains—there could well have been a hylozoic alternative to what we know as Boyle's law.

In the context of mid-1600s England, Robert Boyle (1627–1691) chose to ground his scientific view on mechanistic principles ("the Mechanical or Corpuscular Philosophy," as he refers to it) and preferred them over hylozoic principles, which were advocated to argue that all matter was animated, and also to challenge the social stability and status quo of the time. Ideas on social, economic, political and gender equality were put forward, and political initiatives organized, to build a fair society. On his side, Robert Boyle extensively wrote on women and their social role, and on what kind of men—men with strong masculine features—could best do science. In his epistolary writings on women, Boyle recommends women be chaste and modest, to aid men's chastity, which he saw as linked to undistracted devotion to God and experimental science. Women are clearly put by Boyle at the margins of experimental science, being in the picture just to shape the portrait of the experimental man. In his research as a chemist, Boyle was animated

[6]For some works on feminist approaches to physics and mathematics, see, e.g., Barad (1996); Shulman (1996).

by the intent to elaborate hypotheses that well accorded with the data, but—Potter insists—socio-cultural constraints, which were reinforcing a traditional feminine figure, played an important part too. Animism was contrasted and rejected by Boyle not simply because it did not accord well with the data, but also to eschew its moral and political—radical and subversive—connotations. Such scientific hypotheses as "nature abhors a vacuum" and "the air has spring and weight" are hence to be read in the historical context of Boyle's work, and as underpinned by some intersection of empirical investigation, religious beliefs, and gender issues. Potter's reconstruction of the sexist elements related to the roots of Boyle's religious and political perspectives includes a very wide range of elements taken from history, history of science, sociology of science, and political history. Boyle's experimental work, collection of data and production of hypotheses stand as an exemplary case of the interplay of epistemic and non-epistemic values in the construction of sound and still accepted scientific knowledge.

Further work has recently been carried out on possible relations between feminist science studies and chemistry by Kovács (2012a, b), who highlights the distance between the philosophical ideas underlying chemical thermodynamics, Boyle's modeling of ideal gas and solutions, and the values promoted by feminist epistemology. Kovács does not refer just to Boyle's religious and political convictions from an historical perspective, but partly appeals to Longino's theory to elaborate a conceptual analysis of Boyle's work. Kovács criticizes Potter's work insofar as it determines the connections between scientific statements and cultural values empirically, establishing the value-ladenness of scientific knowledge contingently on whether the historical actors involved perceived it as value-laden and were aware of their interests. Longino's contextual empiricism, instead, recognizes the role of background assumptions and contextual values no matter whether they are consciously or unconsciously assumed by scientists relying on them. In this perspective, the origins of assumptions and values and their relations with evidential claims are located in the belief system of the researchers' culture, were the individual scientist aware of them or not.

Kovács' main concern is to clarify what can count as gender ideology in the context of chemical theories, which deal with inanimate nature, and identifies it with a general philosophical framework. The holding of atomism, the assumption of a rigid hierarchy among states of matter, and the disregard of interrelationships among parts in investigations of ideal gases are taken as guided by gender ideology.[7] Not only do such positions neglect feminist epistemological values, but they do so—it is argued—at the expense of a representation of gases that might be closer to real systems.[8]

> There is a scale of perfection-imperfection on which states of matter are positioned, with the ideal gas on the top, real gases in the middle and liquids at the bottom. [...] Ultimately, the

---

[7]"Sexism is a special case of hierarchical thinking" (Kovács 2012b: 123).

[8]Kovács (2012a: 106–10) points out how intermolecular forces, which are essential in *real* gases, are neglected in the treatment of *ideal* gases.

> scale of perfection/imperfection is a function of the intermolecular forces in the system and of the extension of the molecular entities in space, i.e. their embodiment. Interactions and embodiment are seen as problems, as deviation from the ideal state (Kovács 2012a: 103).

Embracing of a different standpoint would have led to a different way of modeling gases.

Once we claim that different modeling could have emerged from different gender considerations, how many other stances should be embodied in the rethinking of our scientific principles, and to what end? What we have presented stands as an important feminist contribution to a historical re-consideration of some core principles of an apparently non-gendered discipline. Which outcomes some extension of the same kind of approach to other fields and contexts could convey should be further investigated.

12.2.3 On Feminist Criminology

As already pointed out, feminist literature has stimulated a substantial reshaping of research in a number of fields by calling attention to biased assumptions and bringing new evidence to the fore. This is the case with feminist criminology, a field that after its start in the early Seventies has been constantly growing ever since.[9] This branch of criminology aims at explaining the gender gap in crime, namely the different rates of criminal behavior among men and women, and female criminal behavior in general. There are a number of different perspectives within feminist criminology, depending on what source of gender inequality is taken to exercise an influence on female criminal behavior. Among the trends identified to date, the *liberal*, *radical*, *Marxist*, and *socialist* feminisms, liberal feminists relate crime to the role of women occupied within society, and explain the lower rate of female criminality as a function of women's lower social status: "women offend at a lower level than men because their socialization provides them with fewer opportunities to engage in deviance" (Burgess-Proctor 2006: 29). Rather than focusing on roles, radical feminists concentrate on the ways in which patriarchy-dominated society favors male crimes against women, such as rape, violence, harassment, and the like. Marxist feminists stress the subordinate state of women within capitalist societies and identify a major source of female criminality with their economic marginalization, which explains why women are more prone to misdemeanor like petty theft than felony. Socialist feminism "combines radical and Marxist perspectives to conclude that women's oppression results from concomitant sex- and class-based inequalities. […] Within criminology, socialist feminists examine causes of crime within the context of interacting gender- and class-based systems of power" (Burgess-Proctor 2006: 29).[10]

---

[9]An overview of the literature on feminist criminology up to 2000 is to be found in Britton (2000).

[10]See also the discussion of these trends in Williams and McShane (2004: Chap. 13).

Alongside studies purporting to explain female criminal behavior, extensive work has been done to embody the gender variable into theories focusing on criminal behavior in general. These studies stress the need to take into account gender in addition to socioeconomic factors such as class, race, employment, income, and so on, in order to understand criminal behavior. One such perspective is the power control theory of gender and delinquency developed by Hagan (1988; see also Hagan 1994), which calls attention to the importance of situating the analysis of crime within the context of power relationships both in the working environment and in the household. Central to this theory is the idea that power and control structures in the workplace are reproduced in the family, resulting in unequal social control of girls and boys. This leads to different attitudes toward risk-taking in patriarchal and liberal families, which is in many ways related to delinquency. Hagan's theory includes an insightful analysis of the inclination to rebel among girls and boys, observing that girls have different ways of reacting to social control. In a similar vein, other authors have investigated criminal behavior in relation to the way in which sex, race and societal status are perceived within different types of social structures.

Other studies extend to gender the theory that strain caused by the inability to obtain success, especially of a monetary kind, is leading to crime—the so-called strain theory—and explore the relationship between strain exposure and criminal behavior from a gender-oriented perspective. This has led to extend the concept of strain to a wider spectrum of factors, such as cognitive, emotional and affective, and to explore the reactions of women and men. The guiding hypothesis here is that female crime is rooted in the strain caused by the oppression women undergo in societal and private life, such as the frustration due to failure to achieve the desired status in the workplace and in society at large, the restriction of behavior imposed on women by cultural and religious conventions, the experience of sexual and emotional abuse. As observed by Lisa Broidy and Robert Agnew, studies conducted in this perspective "pull together and synthesize the theory and research from several areas, providing a more precise statement of those variables that may be important in explaining gender differences in crime and the causes of female crime. Further, they point to a range of new variables for empirical researchers to consider" (Broidy and Agnew 1997: 298).

Although fragmentary, the preceding remarks provide evidence that investigation into gender and crime has highlighted unexplored aspects of criminal behavior in general, promoted the search for novel data and opened new directions of research.

12.3 Philosophy of Science and Social Responsibility

Feminist work started in a critical vein, moving from analyses of women's roles in the sciences and women's ways of knowing, to a constructive reframing of epistemological issues, aiming at developing an account of scientific knowledge that adequately considers the context and status of knowers. The identification of cognitive authority "is usually associated with a cluster of markings [...]. Gender identity cannot be adequately understood—or even perceived—except as a component of complex interrelationships with other systems of identification and hierarchy" (Alcoff and Potter 1993: 3). Socio-political elements deeply participate in such complex interrelationships, and it is hence worth considering whether the concerns expressed by feminist epistemologists are to be confined to a strictly scientific and academic context, or rather related—and how—to wider socio-economic and political stances. Different positions on this matter can also affect views over whether the numerous dimensions of feminist epistemologies do, can, or should impact on the overall role of philosophy of science as a discipline.

A whole range of interrelated questions have been arising from within feminist studies regarding the reasons better paid occupations with a higher status are still largely held by males and why science is still commonly perceived as "masculine" and family care (e.g. caring for children and the elderly) as "feminine," within an unbalanced socio-economic environment. The underrepresentation of women in science has been shown to be not the consequence of some objective, innate lesser ability or unsuitability, but one of the results of "a complex mesh of social and economic forces excluding women from high status occupations generally, reinforced by the influence of false and confused ideas about how women's cognitive abilities differ from men's and about what the demands of scientific method are" (Haack 1992: 12). Is philosophy of science supposed to contribute to combating such forces in order to promote a more balanced perspective on cognitive activities and, in the end, a non-discriminating social context?[11]

Gender-sensitive analyses of scientific inquiry elaborated by feminist epistemologists stress the social and dynamic dimensions of the production of scientific knowledge, the role of factors such as trust and credibility, patterns of communication and distribution of research efforts among scientists. The denunciation of practices of marginalization of women (and other minorities) by a group of dominant epistemic subjects (usually male) led feminist epistemologists to focus on the need to enhance inclusive and constructive dialogue and democratic exchange. Given the situated nature of epistemic agents, contexts and values should be incorporated as integral to the acquisition of knowledge, and social accounts of objectivity be set in social accounts of scientific knowledge. As recalled above, Longino's contextual empiricism has played a major role in the elucidation of these topics, and in the promotion of a genuinely democratic scientific community. We regard her account as particularly convincing and promising in eschewing the concept of a single truth or a single adequate epistemological perspective, recognizing the relevance of background assumptions, and suggesting that knowledge be produced by inclusive scientific communities, in which views and criticisms are democratically exchanged (Longino 1990, 2002, 2008). In this perspective, which has clear social and political implications, representations of phenomena are not evaluated just with respect to what is

[11]See VV.AA. (2013).

represented, but also with respect to the cognitive needs and values of the community elaborating them, which ought to be made explicit.

Shall the future course of philosophy of science be significantly influenced by the development of feminist discourses on science and interactions with social values, and, if so, how and to what extent? Roughly a decade ago, Kourany (2003: 3) argued that "a new program for philosophy of science" was to be promoted, according to which philosophy of science should cover a socially responsible role. Feminist science studies have been struggling for quite some time now for women's equality —both as objects and subjects of the scientific enterprise—helping to overcome prejudice and discrimination within the sciences, and shape an egalitarian perspective on women's potentialities and accomplishments on a large scale. Must the overcoming of gender biases in wide socio-political contexts be one of the goals of the "philosophy of science for the twenty-first century," as Kourany hopes? Should some sense of social responsibility and participation in the politics of knowledge belong to the *disciplinary specific features* of philosophy of science?

Kourany points out how concerns of this kind were strongly present at the very roots of the discipline: for the members of the Vienna Circle science was to take part in a reform of society, and philosophy of science was to contribute to such a reform, with the elaboration of a scientific world conception meant to penetrate personal and public, cultural and social life.[12] The development of the field, with the success of Anglo-American empiricism, is held to have somehow betrayed part of the original spirit by promoting a socially disengaged philosophy of science. Were we to agree with Kourany's view, the work of feminist epistemologists should extend to politically engaging tasks such as, for instance, health care, regulation of the labor market, and the educational establishment. Following Kourany's line of reasoning, a socially and politically committed philosophy of science would be a philosophy of science truer to its vocation and role, having to do also with the promotion of egalitarian ideals. Would a stronger embracing of such ideals improve the overall project of philosophy of science, and respond to its specific targets as a disciplinary field?

Kourany (2003: 10) provides some very practical reasons to support the feminist project, claiming it should be promoted

> because society—both women and men—ultimately pays for science. And because society is deeply affected by science. [...] Surely one of the needs of society—both women and men—is justice, and equality for women is one aspect of that justice. This is why science's set of empirically acceptable options should be narrowed to include only the ones that support (or most support) egalitarian goals. This is why the feminist project of appraising science in terms of an egalitarian ideal of human flourishing should be pursued.

It is arguable whether these motives suffice to make philosophy of science converge on the proposal to make the feminist project become a permanent, constituting part of philosophy of science, and to drive philosophy of science itself

---

[12]On the Vienna Circle movement as a form of political engagement see Romizi (2012), Uebel (2005).

towards a socially responsible way to pursue philosophical inquiries. Doubts in these respects have been raised by Giere (2003), who stresses how the program of including pragmatic and social features in our understanding science has been largely pursued also by historians and sociologists of science, and is already quite well-established. Kourany's project to turn the philosopher of science into an active political subject is far from finding consensus in philosophy of science. As Giere points out, political activities are not part and parcel of the professional lives of philosophers of science, active political engagement does not normally count in terms of academic scientific outcomes, it rather subtracts time that could be devoted to their achievement. Kourany's ideal of a politically active philosophy of science might run the risk of driving it away from its genuine pursuit as a professional discipline, instead of fulfilling its original vocation as she suggests.

Is there any less controversial lesson philosophy of science can take from feminist epistemologies? A few relevant feminist suggestions seem to be accompanied by problematic issues. As we have seen, feminists stress how epistemic agents are *communities*, not single individuals. Actual scientific work is performed through interactions between scientists inside and outside the laboratory, and between scientists and funders, industries, public opinion, and users of scientific results. Epistemic communities can be more or less homogeneous, more or less internally diversified, and hence a given set of beliefs and assumptions about gender, race, class, sexuality, and other socio-political categories can be more or less widespread and shared within the group. The more homogeneous an epistemic community is, the more widely some assumptions are to be held, and the more likely they are to go unnoticed. If interests and values are unlikely to be revealed in a homogeneous scientific community, it seems we should wish epistemic heterogeneity be maintained, and possibly enhanced. Philosophical work should, amongst others, help reveal hidden biases, make them explicit, and identify marginalized knowers. At the same time, from a feminist perspective more democratic knowledge procedures should be promoted to incorporate non dominant standpoints (see e.g. Harding 1993). A tension then seems to emerge, on the one hand, between the important role of epistemic diversity to ensure that assumptions, beliefs and values are brought to light, and on the other the purpose of promoting a more inclusive and democratic science, in which women's—and other marginalized—standpoints are included and integrated. Donna Haraway states: "feminism is about a critical vision consequent upon a critical positioning in in-homogeneous gendered social space" (Haraway 1988: 181). The social space is in-homogeneous not only due to gender differences, but for a number of other reasons as well, such as economic, ethnic, class, religious, and other features. A genuine overcoming of traditional ways of conceiving scientific knowledge and its production has to avoid both discrimination between dominant and marginalized groups of knowers and sectors, and a reduction to indifference between various epistemic standpoints. Which kinds of in-homogeneity and how much epistemic diversity it is optimal to have remain crucial open issues.

Feminist reflections strongly point out the impossibility of an "innocent," "from nowhere" and detached epistemology, and stress the need for strategies that include the vantage point of the subjugated—women in the first place. If knowledge is and always will be the result of dominant positions, and feminist thought aims at enhancing the role of women in the sciences, attention must be paid not to end up reinforcing oppositions or dualisms, by finally having women in a dominant epistemic position versus some other subordinated epistemic group. On the one hand, the presence of epistemic minorities and outsiders is presented as revealing interests, assumptions and values; on the other, the goal is to thoroughly integrate minorities into scientific communities.[13] A tension is likely to arise in this respect as well. "Without epistemic privilege, there appears no way to decide between or among conflicting knowledge claims produced by different groups of women." According to Harding's feminist standpoint theory, "different groups of women (among other groups) have not epistemic privilege, but epistemic advantages arising from their distinctive understandings of the world" (Potter 2006: 155–156). The recognition of some sort of epistemic privilege or advantage from a distinctive position—the difference isn't entirely clear and convincing—seems to be necessary to decide between competing knowledge claims. Overcoming marginalized positions, though, should lead to the recognition of all epistemic positions being endowed with the same epistemic authority.

Although with the problematic aspects just highlighted, feminist philosophy of science sheds light on important aspects of how scientific knowledge is actually produced. As remarked above, this is by no means a project belonging exclusively to feminist philosophy, and it does not directly and necessarily commit philosophy of science to specific responsibilities at a socio-political level. However, it can be agreed that "a project for the next generation of feminist philosophers of science is to continue to explore, develop, and articulate philosophical frameworks for modeling the interaction between gender ideologies and science—*in ways inclusive of, but not restricted to, the question of bias*." In order to do so, they need to "offer an alternative, multidisciplinary, non-reductionist and multilayered view of how we might understand sex and gender in all of its dimensions" (Richardson 2010: 357–358, italics added), as well as other sources of bias and kinds of assumptions, as effectively emphasized by Longino. Precisely locating knowledge claims into a possible web of connections, and analyzing them critically, are thus the responsibilities philosophy of science shall take charge of, fostering interactions and transformations of systems of partial perspectives.[14]

Feminist philosophical work on the engendering of scientific knowledge and on epistemic power contributes by unmasking how scientific knowledge is not value-neutral, and unraveling evaluative standards, competing interests and rival answers to the same questions. The responsibility philosophy of science is taking

---

[13]Let us also recall that there is per se no single "women's standpoint." Women can in turn differ by culture, race, ethnicity, class, and other features. The very definition of what exactly counts as the feminist standpoint is hence worth deeper reflection, as feminist empiricist standpoint theorists recognize.

[14]For some reflections regarding why this position does not imply relativism, see e.g. Haraway (1988: 178–179).

upon itself is to offer adequate analyses of background assumptions and contextual social values, and their relations with evidential considerations.[15] Objectivity thus emerges as the result of intersubjective criticism and constant critical questioning of received views and beliefs. "While it is not irrational for scientists to use contextual values in making decisions among competing hypotheses etc., it is irrational for them to do so un-self-consciously, without careful thought. [...] In the same way, it is not irrational for philosophers to use contextual values in producing or supporting their philosophical accounts. But it is irrational to do so without careful thought or without any awareness of doing so" (Potter 2006: 176). The responsibility of philosophers of science to the scientific community, and, in the end, the whole society, can hence rest in the creation of *critical* epistemic agents, aware that neither they nor scientists can pass as neutral through contextual values and assumptions, and in being ready—as feminist philosophy of science has taught them—to "wear their central contextual values on their philosophical sleeves" (Potter 2006: 176).

References

Alcoff, L., & Potter, E. (1993). Introduction: When feminisms intersect epistemology. In L. Alcoff & E. Potter (Eds.), *Feminist epistemologies* (pp. 1–14). London: Routledge.

Anderson, E. (2011). Feminist epistemology and philosophy of science. In E. N. Zalta (Ed.), *The Stanford encyclopedia of philosophy*. http://plato.stanford.edu/archives/fall2012/entries/feminism-epistemology/. Accessed January 15, 2015.

Barad, K. (1996). Meeting the universe halfway: Realism and social constructivism without contradiction. In L. H. Nelson & J. Nelson (Eds.), *Feminism, science, and the philosophy of science* (pp. 161–194). Dordrecht: Kluwer.

Britton, D. M. (2000). Feminism and criminology: Engendering the outlaw. *Annals of the American Academy of Political and Social Science, 571*, 57–76.

Broidy, L., & Agnew, R. (1997). Gender and crime: A general strain theory perspective. *Journal of Research in Crime and Delinquency, 34*, 275–306.

Burgess-Proctor, A. (2006). Intersections of race, class, gender, and crime: Future directions for feminist criminology. *Feminist Criminology, 1*, 27–47.

Campaner, R., & Galavotti, M. C. (2007). Plurality in causality. In P. Machamer & G. Wolters (Eds.), *Thinking about causes. From Greek philosophy to modern physics* (pp. 178–199). Pittsburgh: University of Pittsburgh Press.

Campaner, R., & Galavotti, M. C. (2012). Evidence and the assessment of causal relations in the health sciences. *International Studies in the Philosophy of Science, 26*, 27–45.

Cellucci, C. (2011). Classifying and justifying inference rules. In C. Cellucci, E. Grosholz, & E. Ippoliti (Eds.), *Logic and knowledge* (pp. 123–148). Newcastle upon Tyne: Cambridge Scholars Publishing.

Cellucci, C. (2013). Philosophy of mathematics: Making a fresh start. *Studies in History and Philosophy of Science, 44*, 32–42.

Craver, C., & Darden, L. (2013). *In search of mechanisms: Discoveries across the life sciences*. Chicago: University of Chicago Press.

Darden, L. (2008). Thinking again about mechanisms. *Philosophy of Science, 75*, 958–969.

[15]On the social engagement of philosophy of science, see Potochnik (Ed.) (2014).

Fausto-Sterling, A. (2000). *Sexing the body: Gender politics and the construction of sexuality*. New York: Basic Books Inc.
Galavotti, M. C. (2006). For an epistemology 'from within'. An introduction to Suppes' work. Epistemologia, *29*, 215–224.
Galavotti, M. C. (2008). Causal pluralism and context. In M. C. Galavotti, R. Scazzieri, & P. Suppes (Eds.), *Reasoning, rationality and probability* (pp. 233–252). Stanford: CSLI.
Galavotti, M. C. (2014). For a bottom-up approach to the philosophy of science. In C. Cozzo & E. Ippoliti (Eds.), *From a heuristic point of view. Essays in honor of Carlo Cellucci* (pp. 199–214). Newcastle upon Tyne: Cambridge Scholars Publishing.
Giere, R. N. (2003). A new program for philosophy of science? *Philosophy of Science, 70*, 15–21.
Haack, S. (1992). Science "from a feminist perspective". *Philosophy, 67*, 5–18.
Hagan, J. (1988). *Structural criminology*. Oxford: Polity Press.
Hagan, J. (1994). *Crime and disrepute*. Thousand Oaks: Pine Forge Press.
Haraway, D. (1986). Primatology is politics by other means. In R. Bleier (Ed.), *Feminist approaches to science* (pp. 77–118). Elmsford, NY: Pergamon Press.
Haraway, D. (1988). Situated knowledges: The science question in feminism and the privilege of partial perspective. *Feminist Studies, 14*, 575–600.
Harding, S. (1993). Rethinking standpoint epistemology: What is 'strong objectivity'? In L. Alcoff & E. Potter (Eds.), *Feminist epistemologies* (pp. 49–82). London: Routledge.
Hennig, C. (2009). A constructivist view of the statistical quantification of evidence. *Constructivist Foundations, 5*, 39–54.
Hennig, C. (2010). Mathematical models and reality: A constructivist perspective. *Foundations of Science, 15*, 29–48.
Keller, E. (1983). *A feeling for the organism*. San Francisco: W.H. Freeman and Co.
Kellert, S., Longino, H., & Waters, K. (Eds.). (2006). *Scientific pluralism*. Minneapolis-London: University of Minnesota Press.
Kourany, J. A. (2003). A philosophy of science for the twenty-first century. *Philosophy of Science, 70*, 1–14.
Kovács, A. (2012a). Gender in the substance of chemistry, Part 1: The ideal gas. *HYLE—International Journal for Philosophy of Chemistry, 18*(2), 95–120.
Kovács, A. (2012b). Gender in the substance of chemistry, Part 2: An agenda for theory. *HYLE—International Journal for Philosophy of Chemistry, 18*(2), 121–143.
Lloyd, E. (1996). Pre-theoretical assumptions in evolutionary explanations of female sexuality. In E. Fox Keller & H. Longino (Eds.), *Feminism and science* (pp. 91–102). Oxford: Oxford University Press.
Longino, H. (1990). *Science as social knowledge*. Princeton: Princeton University Press.
Longino, H. (2002). *The fate of knowledge*. Princeton: Princeton University Press.
Longino, H. (2006). Philosophy of science after the social turn. In M. C. Galavotti (Ed.), *Cambridge and Vienna. Frank P. Ramsey and the Vienna Circle* (pp. 167–177). Dordrecht: Springer.
Longino, H. (2008). Values, heuristics, and the politics of knowledge. In D. Howard, J. Kourany, & M. Carrier (Eds.), *The challenge of the social and the pressure of practice* (pp. 68–86). Pittsburgh: University of Pittsburgh Press.
Martin, E. (1991). The egg and the sperm. *Signs: Journal of Women in Culture and Society, 16*, 485–501.
Potochnik, A. (2012). Feminist implications of model-based science. *Studies in History and Philosophy of Science, 43*, 383–389.
Potochnik, A. (Ed.). (2014). *Socially engaged philosophy of science*. *Erkenntnis*, 79, Supplement 5, 901–1018.
Potter, E. (2001). *Gender and Boyle's law of gases*. Bloomington: Indiana University Press.
Potter, E. (2006). *Feminism and philosophy of science. An introduction*. London: Routledge.
Richardson, S. S. (2010). Feminist philosophy of science: History, contributions, and challenges. *Synthese, 177*, 337–362.

Romizi, D. (2012). The Vienna Circle's "scientific world conception": Philosophy of science in the political arena. *HOPOS, 2*, 205–242.
Schaffner, K. (1993). *Discovery and explanation in biology and medicine*. Chicago: The University of Chicago Press.
Shulman, B. (1996). What if we change our axioms? A feminist inquiry into the foundations of mathematics. *Configurations, 3*, 427–451.
Sperling, S. (1991). Baboons with briefcases: Feminism, functionalism and sociobiology in the evolution of primate gender. *Signs: Journal of Women in Culture and Society, 17*, 1–27.
Suppe, F. (2000). Understanding scientific theories: An assessment of developments, 1969–1998. In D. A. Howard (Ed.), *PSA 1998. Philosophy of Science* (pp. S102–S115). Supplement to Vol. 67(3).
Suppes, P. (1962). Models of data. In E. Nagel, P. Suppes, & A. Tarski (Eds.), *Logic, methodology and philosophy of science* (pp. 252–261). Stanford: Stanford University Press.
Suppes, P. (1970). *A probabilistic theory of causality*. Amsterdam: North-Holland.
Suppes, P. (1981). The plurality of science. In P. D. Asquith, & I. Hacking, (Eds.), *PSA 1978*, Vol. II (pp. 3–16). East Lansing: Philosophy of Science Association. Reprinted in Suppes (1993), pp. 41–54.
Suppes, P. (1984). *Probabilistic metaphysics*. Oxford: Blackwell.
Suppes, P. (1990). On deriving models in the social sciences. *Journal of Mathematical and Computer Modelling, 14*, 21–28. Reprinted in Suppes (1993), pp. 435–450.
Suppes, P. (1993). *Models and methods in the philosophy of science: Selected essays*. Dordrecht-Boston: Kluwer.
Suppes, P. (2002). *Representation and invariance of scientific structures*. Stanford: CSLI Publications.
Uebel, T. (2005). Political philosophy of science in logical empiricism: The left Vienna Circle. *Studies in History and Philosophy of Science, Part A, 36*, 754–773.
van Fraassen, B. (1980). *The scientific image*. Oxford: Oxford University Press.
Van Regenmortel, M. H. V., & Hull, D. L. (Eds.). (2002). *Promises and limits of reductionism in the biomedical sciences*. Chichester: Wiley and Sons.
VV.AA. (2013). Forum: Pluralism? Reflections on women in the history of philosophy of science. *HOPOS, 3*, 135–150.
Williams, F. P. III, , & McShane, M. (2004). *Criminological theory* (4th ed.). Upper Saddle River: Pearson Education.
Wimsatt, W. (1974). Complexity and organization. In K. Schaffner & R. S. Cohen (Eds.), *PSA-1972, Boston studies in the philosophy of science* (pp. 67–86). Dordrecht: Reidel.
Wimsatt, W. (1994). The ontology of complex systems: Levels of organization, perspectives, and causal thickets. *Canadian Journal of Philosophy, 20*, 207–274.
Wimsatt, W. (2006). Reductionism and its heuristics: Making methodological reductionism honest. *Synthèse, 151*, 445–475.

MIX
Papier aus verantwortungsvollen Quellen
Paper from responsible sources
FSC® C105338
FSC
www.fsc.org

If you have any concerns about our products,
you can contact us on
ProductSafety@springernature.com

In case Publisher is established outside the EU,
the EU authorized representative is:
Springer Nature Customer Service Center GmbH
Europaplatz 3, 69115 Heidelberg, Germany

Printed by Libri Plureos GmbH
in Hamburg, Germany